Reversing Climate Change

How Carbon Removals can Resolve Climate Change and Fix the Economy

Reversing Climate Change

How Carbon Removals can Resolve Climate Change and Fix the Economy

Graciela Chichilnisky
Columbia University, USA

Peter Bal
Millemont Institute, USA

World Scientific

NEW JERSEY • LONDON • SINGAPORE • BEIJING • SHANGHAI • HONG KONG • TAIPEI • CHENNAI • TOKYO

Published by

World Scientific Publishing Co. Pte. Ltd.
5 Toh Tuck Link, Singapore 596224
USA office: 27 Warren Street, Suite 401-402, Hackensack, NJ 07601
UK office: 57 Shelton Street, Covent Garden, London WC2H 9HE

Library of Congress Cataloging-in-Publication Data
Names: Chichilnisky, Graciela, author. | Bal, Peter (Ecologist) | Conference of the Parties (United Nations Framework Convention on Climate Change) (21st : 2015 : Paris, France)
Title: Reversing climate change : how carbon removals can resolve climate change and fix the economy / Graciela Chichilnisky, (Columbia University, USA), Peter Bal, (Millemont Institute, USA).
Description: New Jersey : World Scientific, 2016. | "This book is about the 21st session of the Conference of the Parties to the United Nations Framework Convention on Climate Change (UNFCCC), or COP21, held from 30 November to 11 December 2015 in Paris"--After table of contents. | Includes bibliographical references.
Identifiers: LCCN 2016035722| ISBN 9789814719346 (hardcover) | ISBN 9789814719353 (pbk.)
Subjects: LCSH: Climate change mitigation--Congresses. | Emissions trading--Congresses. | Carbon dioxide mitigation--Congresses. | Greenhouse effect, Atmospheric--Congresses. | United Nations Framework Convention on Climate Change (1992 May 9). Protocols, etc. (1997 December 11)
Classification: LCC TD171.75 .C45 2016 | DDC 363.738/746--dc23
LC record available at https://lccn.loc.gov/2016035722

British Library Cataloguing-in-Publication Data
A catalogue record for this book is available from the British Library.

For any available supplementary material, please visit
https://www.worldscientific.com/worldscibooks/10.1142/9765#t=suppl

Desk Editor: Sylvia Koh

Typeset by Stallion Press
Email: enquiries@stallionpress.com

Dedication

To Natasha Sable Chichilnisky-Heal

Genius and Grace

Praise for the Authors and the Book

Avoiding the immense dangers of unmanaged climate change requires holding global temperature increases to the Paris COP21 target of "well below 2 degrees Centigrade." Currently, future likely or planned emissions paths fall far short of the reductions necessary to achieve this goal. Strong and urgent reductions of emissions are vital and it is increasingly clear that we will also need to remove carbon dioxide from the atmosphere. To incentivise efficient and effective action on the scale now necessary, we will need new technology and a strong carbon price to reward reductions and removals. This important book not only makes these arguments clearly and strongly, it also argues that the necessary technologies for removal, including those developed by the authors, are now available.

Nicholas Stern

IG Patel Chair of Economics and Government,
Department of Economics, London School of Economics

The genius of Graciela Chichilnisky is recognized by economists and with this book she has focused that talent to the dire problem facing mankind. To survive we must do more than stave off a further rise of CO_2 in the atmosphere. We need to reverse it if the planet is to be viable. Professor Chichilnisky's achievement along with her co-author Peter Bal is to show us the way to rescue our future.

Professor Edmund Phelps

2006 Nobel Laureate in Economics
Director, Center on Capitalism and Society,
Columbia University

The team of Chichilnisky and Bal has exceptional skill in explaining complex topics with great clarity making it easy for non-scientists interested in climate change to read. They address the science of climate change, the complex international negotiations needed to reach a compromise between developing nations and the developed ones, and importantly the urgent need to find a way of extracting CO_2 from the atmosphere and utilizing and sequestering it in a commercially profitable manner. The last topic has been almost completely ignored by the media.

Theodore Roosevelt IV

Managing Director & Chairman of Barclays Cleantech Initiative
BARCLAYS

Graciela Chichilnisky's leadership in the global community, in producing policies to reduce the greenhouse gas emissions that cause global climate change, has been revolutionary.

Jay Inslee

Former Member of the United States House of Representatives; Governor of Washington.

In the world of economic theory, Graciela Chichilnisky is an A-list star.

The Washington Post

Graciela Chichilnisky has made important contributions to the economic aspects of climate change, both in analysis and in the formulation of appropriate policies. In particular, she has emphasized the considerations of justice in a manner capable of reasoned analysis.

Kenneth Arrow

Winner of the Nobel Memorial Prize in Economic Sciences

Graciela Chichilnisky is one of the most incisive minds working on the subject of justice among the generations.

Sir Partha Dasgupta

Frank Ramsey Professor of Economics at Cambridge University

About the Authors

Graciela Chichilnisky worked extensively on the Kyoto Protocol, creating and designing the carbon market that became international law in 2005. In 2017, she was selected by the Carnegie Foundation for their prestigious Great Immigrant, Great American award showcased in the *New York Times*, and in 2018 she was awarded the Albert Nelson Marquis Lifetime Achievement featured in the *Wall Street Journal*. *The Washington Post* calls her an "A-List Star" and *Time Magazine* calls her a "Hero of the Environment." U.S. Congressman Jay Inslee wrote that her work is "revolutionary for the international community." A world-renowned economist, she is the creator of the formal theory of Sustainable Development and acted as Lead U.S. Author of the Intergovernmental Panel on Climate Change, which received the Nobel Prize in 2007. Her pioneering work uses innovative market mechanisms to create Green Capitalism. Dr. Chichilnisky acts as a special adviser to several U.N. organizations and heads of state.

Dr. Chichilnisky is CEO and Co-Founder of Global Thermostat (www.globalthermostat.com), a company that has created a "Carbon Negative Technology"™ that captures CO_2 from air and transforms it into profitable assets, including biofuels, food, beverages, and enhanced oil recovery. Earlier, she founded and led two successful companies: FITEL, a financial telecommunications company that was sold in Japan;

and Cross Border Exchange, a global technology communications company sold to J. P. Morgan.

Additionally, Dr. Chichilnisky is a Professor of Economics and Mathematical Statistics and a University Senator at Columbia University, and Director of the Columbia Consortium for Risk Management (www.columbiariskmanagement.net), where she developed a landmark methodology, with support from the U.S. Air Force, for a new foundation of probability and statistics in an approach to catastrophic risks that allows more realistic treatment of rare but important events. She is the author of 15 books and some 320 scientific articles in preeminent academic journals. Her two most recent books are *The Economics of the Global Environment: Catastrophic Risks in Theory and Policy,* and *Manifolds of Preferences and Equilibrium.*

Dr Chichilnisky holds two PhD degrees, in Mathematics and Economics from MIT and UC Berkeley. She is a frequent political and economic commentator on *CNN*, *ABC*, *BBC TV News*, and *Bloomberg News*, as well as a frequent keynote speaker at leading international conferences and universities. She taught previously at Harvard, Essex and Stanford Universities, appeared in *Time Magazine* on "Heroes of the Environment," and was elected one of the Ten Most Influential Latinos in the U.S.

Dr. Chichilnisky is currently a Visiting Professor at Stanford University. *Fast Company* selected her company, Global Thermostat, as a "World's Top Ten Most Innovative Company" in Energy. Dr. Chichilnisky was selected as the 2015 "CEO of the Year" by IAIR, a title awarded at the Yale University Club in New York City. In 2019, MIT Technology Review chose the carbon removal technology that she co-authors as "Top Ten Breakthrough Technologies" in the world, an award curated by Bill Gates.

Peter Bal is a businessman and ecological restoration practitioner. Born in the U.S. in 1960 and educated across Europe and Asia, Bal uses his multilingual and creative mind to bridge cultural differences in international transactions. His experience includes selling the Empire State Building, producing alcohol from grapes and restoring a devastated forest in France.

Bal sees CO_2 as an asset to be mined. He is dedicated to natural plant absorption, as well as to industrial solutions for retrieving CO_2 from the atmosphere. He is currently working on a containerised CO_2 absorbing unit with Global Thermostat and setting up ecological research and training centres with John D. Liu.

Contents

Introduction: Climate Change and our Future

I am drinking and eating my way through pastures, trees, minerals, rivers, oceans, and other species. I am polluting and changing the planet's atmosphere and its oceans, and destroying the myriad species that comprise life on earth. In doing so, I am endangering my own life support systems, as I need air, water, and food to survive. Although I am capable of creating innovative ideas, and though technologies and solutions are available and known, I continue to participate in the destruction of our planetary systems, the Garden of Eden. I know the time has come to do something and declare that I shall do what it takes. This is the point of no return. May my actions help create a livable world for everyone.

195 nations negotiated a global climate pact in Paris at COP21, in December 2015. This book is about the challenges and the opportunities created by the climate crisis. It explores the irreversible dangers that we face, and the opportunities available to us now. We are at a critical point; although there exist today solutions to avert climate change, the next few years will determine whether we will implement these solutions or unleash irreversible and, possibly catastrophic, damage. We are truly at a point of no return.

In 1997, the poor and rich nations of planet Earth reached an international agreement to forestall a global disaster: the Kyoto Protocol of the United Nations. It became international law in 2005. For the first time, the Kyoto Protocol capped the emissions of the main emitters, the industrialized countries, one by one. It also created an innovative financial mechanism, the Carbon Market (now the EU ETS) and its Clean Development Mechanism (CDM), which allows developing nations to receive carbon credits when they reduce their emissions below their baselines. The carbon market, an economic system that created a price for carbon for the first time, is now used in four continents, is promoted by the World Bank, and is recommended by leading oil and gas companies.[1] On September 24, 2015, China officially adopted the carbon market, which made global news. The CDM has the potential both to reduce the emissions responsible for climate change and to transfer wealth and clean technology to poor countries to implement the solutions that exist today. The CDM has transferred billions of dollars to poor nations for clean projects since 2005. The essence of the CDM is creating value for all, while decreasing global emissions of CO_2.

The Kyoto Protocol is a historic agreement, the first of its kind. It limits global emissions and creates a new market that is based on trade in user rights to the atmosphere — the global commons — which we all share. The Kyoto Protocol is a work in progress that has taken more than 20 years to build and improve; it did succeed in getting almost all nations to cooperate to reduce global emissions and has achieved about 22.6% in the reduction of emissions since 2005 by the nations with binding commitments.[2] This is a good beginning, but there is still a lot to be achieved.

First, we need to understand how we dug ourselves into this hole in order to figure out how to get out, a topic we will look at in more detail in the first chapter. The inception of industrialization two centuries ago dramatically expanded and accelerated after World War II, and depended heavily on burning fossil fuels — coal, oil, and natural gas — to provide energy to the economy. Energy is the mother of all markets. Everything is made with energy. Economic development still depends on the availability of cheap energy sources, and in today's global economy, this still means fossil fuels. Fossil fuels generate roughly 70% of the electricity in the world today.[3]

The consequences of our thirst for fossil fuels have become increasingly apparent. The science is new and there are still uncertainties, but the risks are real and solutions exist today. The Arctic Sea and glaciers are melting before our eyes. Thawing permafrost in high-latitude and high-elevation regions is causing enormous damage to infrastructure and is rapidly changing ecosystems.[4] According to the World Health Organization (WHO), more than 150,000 people die and millions more become ill each year because of climate change.[5] Ominous signs of a changing climate abound. Heat waves in Western Europe claimed 70,000 lives in 2003, and monsoons left 24% of Bangladesh underwater in 2004. We have seen an increase in the average intensity of Atlantic hurricanes, such as Hurricanes Katrina, Sandy, and Maria, which wiped out much of the U.S.'s Louisiana and Mississippi Gulf Coast shorelines in 2005, closed down most of New York City for weeks in 2011, and destroyed much of Puerto Rico and the Caribbean in 2017.[6] In the last decade, Australia suffered record droughts and widespread fires, followed by record breaking floods, and significant parts of Greenland's ice sheet and the Antarctic continent have already melted.[7] The North and South Poles are rapidly melting. The Pentagon identifies climate change among its top security risks.[8]

A Toxic Dependence

Despite this, our demand for fossil fuels continues unabated. China is a world environmental leader, but, according to U.S.

government projections, it will build a new 600 megawatt plant every 10 days for 10 years.[9] The average U.S. consumer uses more energy today than ever before, despite advances in energy efficiency, partly because the U.S. has some of the lowest oil prices in history. Prices are heavily dependent on international market conditions and on the trade policies of some developing nations. Today's low prices follow a period when oil prices reached some of the highest levels since the Organization of the Petroleum Exporting Countries (OPEC) oil embargoes of the 1970s. The desire for energy independence has created powerful incentives for countries, including China and the U.S., to use their abundant coal and tar sands resources to meet their rapidly growing energy needs.[10] This is bad news, as coal and tar sands are the worst fossil fuels in terms of the amount of carbon they emit.

Contrary to common wisdom, there are solutions today for the climate crisis, even though no silver bullets exist to rid us of our fossil fuel dependence in the short run. There are short-term solutions that involve new carbon capture technologies. These reduce the CO_2 that is already in the atmosphere, as will be explained below. Long-run solutions are different, and they involve a transformation of the energy infrastructure of the world, which is worth about $55 trillion. A former Executive Director of the International Energy Agency (IEA), Nobuo Tanaka, believes we need an energy revolution. Tanaka said in 2008 that we need to change the world's energy infrastructure at a cost of $45 trillion by 2050, which is two-thirds of the gross domestic product (GDP) of the entire planet.[11] Updated figures from 2014 show that it would take $53 trillion by 2035 to get the world on a path to limiting warming by 2°C.[12] It is safe to say that this change will not happen quickly. It is a race against time.

New technologies can rid the atmosphere of its CO_2, and some are starting to be used commercially. Called *carbon negative technologies*™, they can absorb the carbon that is already in the atmosphere and make this carbon into useful commercial products and services that are sold at a profit. The Intergovernmental Panel on Climate Change (IPCC) Fifth Assessment Report of 2014 and 2018 documents that removing carbon from the atmosphere is the only solution in most scenarios to avert catastrophic climate change, since we have procrastinated so much that reducing current emissions no longer averts the risk of catastrophic climate change.[13] Once emitted, CO_2 can stay for hundreds of years in the atmosphere. We are already well into levels that can unleash catastrophic climate change.

What is clear is that our grandchildren will inherit a world that is very different from our own. They will either inherit a planet with a severely diminished capacity to support human life or they will inherit a global economic system fueled by cleaner, renewable energy sources that respects ecological limits and is capable of meeting the basic needs of every woman, man, and child. Which path we set them on is entirely dependent on our response to this crisis. So much hinges on the choices we make now.

We have the chance to transform this crisis into an opportunity for renovating the global energy infrastructure. The clean energy industry is growing rapidly. Investment in clean energy jumped 16% between 2013 and 2014 and current investment in this emerging industry exceeds $310 billion.[14] This is hopeful. Perhaps we can turn the crisis into a success story of human ingenuity and cooperation. But let us not get ahead of ourselves. Right now the crisis is worsening and time is running out.

One critical problem for the future is the continuing impasse between the rich and the poor nations. The U.S. has refused, so far, to ratify the Kyoto Protocol unless

China, India, and other poor nations agree to limit emissions. On July 25, 1997, the U.S. Senate passed by a vote of 95 to 0 the Byrd–Hagel Resolution, declaring that the U.S. would not mandate commitments to limit or reduce its greenhouse gas emission unless there are commitments to limit or reduce greenhouse emissions for Developing Country Parties, and the limits do not harm the U.S. economy. Unless the two world's largest greenhouse gas emitters, U.S. and China, limit their emissions, preventing catastrophic climate change does not seem to be a possibility. How will the global community overcome this divide and forge cooperation between rich and poor countries?[15]

In a nutshell, the question comes down to this: who should reduce emissions — the rich or the poor countries?

The rich nations depended heavily on fossil fuels to grow while the rest of the world were left behind. Today, developing and developed countries contribute approximately equal shares of cumulative greenhouse gas emissions for the period of 1850–2010, while developing nations house 83.3% of the world population.[16] Now, poor countries say that it is their turn to industrialize. Can we really ask poor countries to sacrifice their development opportunities now to atone for the rich nations' past sins? Can we suddenly change the rules of industrial development by requiring countries to find a non-fossil fuel dependent path to development? We can try, but we may very well fail. Rich countries, and in particular the U.S., have little credibility at this stage, and many developing nations, such as India, China, and Brazil, are beginning to flex their economic muscles. Accompanying this is a rapid increase in energy use and carbon emissions.

There is a more basic issue at stake. The developing nations do not emit enough today to resolve the problem by

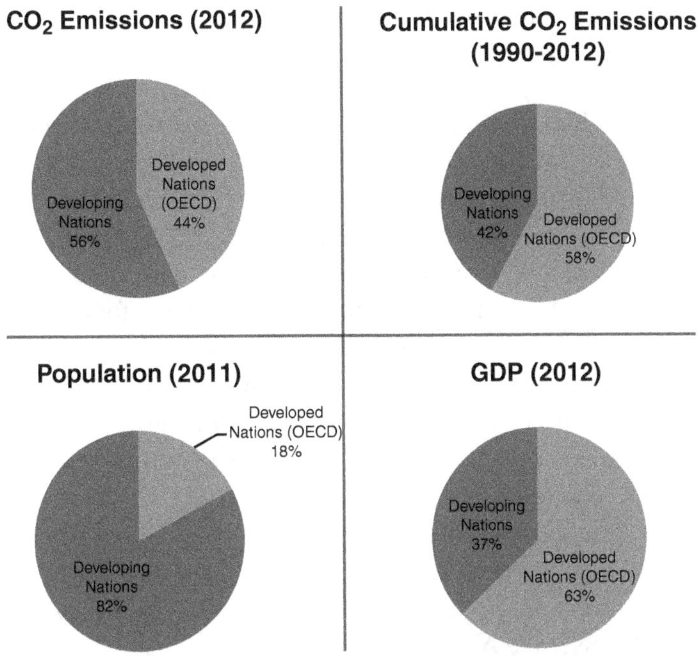

Figure 1 Distribution of world wealth, population, and emissions
Source: OECD: https://stats.oecd.org; The World Bank: http://data.worldbank.org.

themselves unless they stop producing emissions altogether and start using carbon negative technologies. Together, the poor nations emit approximately half of the world's emissions (see Figure 1 "Distribution of World Wealth, Population and Emissions"). Africa, for example, emits only 3% of the global emissions; South America approximately 5.5%.[17] Even if every woman, man, and child in the developing nations, all 5 billion people, stopped emitting any carbon today to oblige the rest — which would require that their citizens stop breathing — they would still not resolve the problem. Developed nations need to reduce global emissions by at least 40%, possibly 70% by 2050.[18] Given who the big emitters are now, the obvious way to achieve this is to decrease emissions in both the industrialized and developing world, working in unison to solve our collective problem.

More importantly, we need to clean up the CO_2 built up in the atmosphere over the last hundred years.

Should the countries that are most responsible for creating the global climate crisis take the lead in solving it? Should the countries that are best able to afford emissions reductions shoulder more of the burden? In 1997, the Kyoto Protocol answered yes to both of these questions and excused developing countries from mandatory emission cuts at that point in time. But times have changed. Developing nations such as China now emit more than the U.S., and the responsibility for global climate change has changed since the Kyoto Protocol was introduced in 1997.[19] It is true that China has four times more people than the U.S. Indeed, China has about 1.4 billion people while the U.S. has about 330 million.[20] Therefore, the average Chinese person is still much more frugal in energy use than the average American.

At the same time, we need to be aware that 20 years from now developing nations could create catastrophic global warming for the entire planet by doing exactly what the rich nations did during their industrialization period: using their own resources to burn fossil fuels. For example, South Africa is the world's fifth largest producer and the sixth largest consumer of coal.[21] This is why the solution to the climate crisis involves all of us. It is truly a one-world problem that nobody can escape.

This book explains why international cooperation between rich and poor nations is crucial in order to prevent catastrophic climate change. The stage for international cooperation has already been set.

The Kyoto Protocol, the only mandatory international treaty to combat climate change, created a carbon market that can unite environmental and economic interests. Its unique properties ensure that our efforts to combat climate change are both fair and efficient. The carbon market — a young market for a global public good — is distinct from any other market in history. That is exactly why it is still a work in progress, even though it has been adopted successfully in four continents. It needs all the help it can get to evolve into a successful solution. Climate change is a complex phenomenon, and it must be recognized that, in the long-run, massive reforestation, hydrological restoration, and a more fundamental respect for life, as a whole, may be needed, topics that go beyond the scope of this book.

This book outlines the history of global climate negotiations that led to the Kyoto Protocol and the Paris Agreement, from the perspective of Graciela Chichilnisky, the architect of the Protocol's carbon market. The Kyoto Protocol carbon emissions limits expire in December 2020, and the 2015 meeting in Paris COP21 did not consider their extension. Many difficult political and economic challenges lie ahead. Although frustration has dominated most of the annual Convention of the Parties, they are still our generation's only uniting force in the struggle against climate change. On June 1, 2017, President Trump decided that the U.S. would withdraw from the 2015 Paris Agreement and, on November 4, 2019, Secretary of State Michael Pompeo began the process to officially withdraw. The withdrawal will take effect one year from delivery of the notification.

1

Global Crisis and the Mandate of COP

Global warming mesmerizes and polarizes public opinion. Climate change is on everyone's minds, yet most people do not fully understand it. A few still deny it. Confusion is understandable. After all, we observe changes in climate from one year to the next. We also know that the earth's climate has changed dramatically over the course of geological history. What is so different now to warrant such alarm?

The difference is that we humans are actively participating in changing the earth's climate in a major way. When scientists refer to climate change, they are talking about changes in the earth's climate above and beyond natural climate variation, which are caused, directly or indirectly, by human activity. The consequence of this activity alters the composition of the earth's atmosphere and causes average surface temperatures on earth to rise. The increase in the earth's surface temperature is known as global warming and is driving climate change. Global warming is now causing polar ice to melt and sea levels to rise. Indeed, sea levels are expected to rise about 10 meters and two to three feet or more by the end of this century according to the U.S. Geological Survey. If this happens, a large part of the world we inhabit today will be under water or vulnerable to severe coastal flooding, including Miami, New York City, Amsterdam, Tokyo, and Shanghai. An analysis of the damage to assets due to flooding estimated the total value to be about \$35 trillion (Table 1, Chapter 2).[1]

How does this happen? When the sun's energy hits the earth, a portion of that energy is reflected back into the atmosphere. Greenhouse gases (GHGs) in the atmosphere, gases such as carbon dioxide, methane, nitrous oxide, carbon monoxide, and others, trap a portion of the heat released by the sun-warmed earth. This is called the greenhouse effect. Although human activity increases other GHGs, carbon dioxide is the gas most responsible for global warming. Carbon emissions comprise more than 80% of the GHGs we produce and remain in the atmosphere for hundreds of years once emitted. For this reason, reducing carbon emissions is the primary focus of global warming prevention efforts. The solution may require using new technology to reduce not just emissions but the actual stock of carbon that is stored in the atmosphere today, a technology known as negative carbon™ and described in the next chapters.

The global carbon cycle is the way the earth balances the carbon exchange between air, oceans, and terrestrial ecosystems (see Figure 1 "The Global Carbon Cycle"). In pre-industrial times, the earth maintained stable carbon dioxide concentrations in the atmosphere and warmed the earth to temperatures that modern human life is accustomed to. These are the temperatures that our bodies and agricultural systems evolved to live in and to thrive on. But the Industrial Revolution marked a turning point: it disrupted the global carbon cycle. Burning fossil fuels, we began pumping carbon dioxide into the atmosphere at a pace faster than the earth could regulate. We began a new era of rapidly increasing atmospheric carbon dioxide levels.

As the Industrial Revolution developed we began our reliance on fossil fuels for industrial, transportation, and

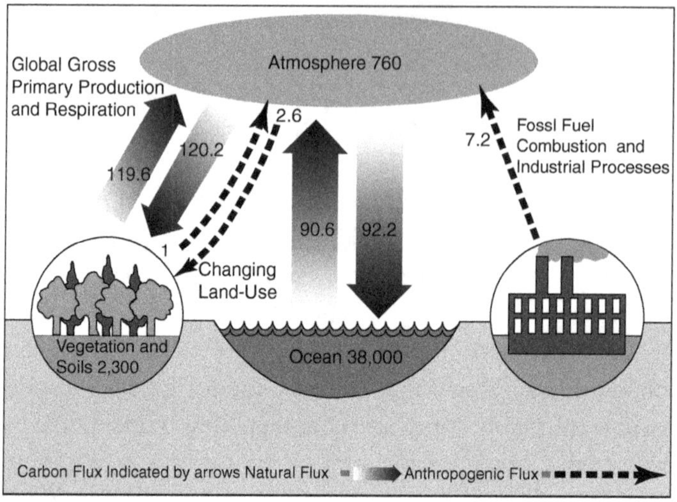

Figure 1 The Global Carbon Cycle (in gigatons of carbon)

home energy uses. This greatly accelerated after World War II with the creation in 1945 of the Bretton Woods Institutions, which include the International Monetary Fund (IMF), the World Bank, and the World Trade Organization, and with the corresponding amplification of global trade and industrial growth that they facilitated. As we burn fossil fuels, such as coal, oil, or natural gas, we release carbon dioxide emissions into the atmosphere. The largest emitters of carbon are the power plants that produce electricity for our homes and our factories. Power plants are responsible for about 45% of all emissions in the U.S. and even more around the world. Transportation is a much smaller but significant contributor of emissions: it accounts for approximately one-third of U.S. emissions and 13% of global emissions. The production, transportation, and distribution of meat itself represent approximately 18–25% of carbon emissions according to the U.N. Of course, this statistic does not include other negative consequences, such as the methane emissions that livestock produce, the pollution of industrial breeding, and the horrible conditions in which meat production takes place.

According to a recent United Nations Food and Agriculture Organization report, *Livestock's Long Shadow*, the various processes involved in bringing meat to the table (meat that is mostly consumed in the rich nations) together emit over 18% of all carbon emissions in the world,[2] more than the entire global transportation system. These emissions intensify the greenhouse effect, increase global temperatures and may cause irreversible damage and rising sea levels.

Forest ecosystems play a critical role in the global carbon cycle. Through photosynthesis, forests sequester carbon dioxide from the atmosphere and store it. The carbon is released back into the atmosphere when trees die and decay. This is all part of the global carbon cycle. But when forests disappear faster than new forests grow, it disrupts the global carbon cycle and contributes to rising atmospheric carbon levels. The conversion of forests to alternative uses, such as when they are harvested for timber products or converted to arable land, releases much of the carbon that had been stored in the wood and the soil back into the atmosphere. This too has contributed to global warming, though to a lesser extent. Scientists attribute roughly 80% of the increase in atmospheric carbon dioxide levels to burning fossil fuels and 20% to land-use change and deforestation. Accounting for emissions from forestry is tricky because when trees die they release more carbon than they captured while alive.

The speed with which we are altering the earth's atmosphere is unprecedented in the history of the planet. Since 1970, GHG emissions have increased by well over 80%, with the 2000s being the highest decade in human history for total GHG emissions.[3] The rate of increase in GHG emissions is now 100 times higher than at the end of the last ice age.[4] If emissions continue to grow at the same rate, atmospheric levels of carbon in 2100 will be double or possibly triple pre-industrial levels.[5] As scientists are quick to warn us, we are heading rapidly into uncharted territory with very little precedent within geological history to guide us. We are driving a global experiment of unknown outcomes with our planet's atmosphere and climate. And we are near the point of no return. The changes we are causing are irreversible, at least for hundreds or thousands of years.

So, what do we know for certain? The overwhelming scientific consensus is that average temperatures on earth are rising due to human causes. This is the assessment of the Intergovernmental Panel on Climate Change (IPCC), an international group of experts formed by the United Nations in 1988 to review the scientific research on climate change, its causes and its anticipated effects. Scientists know that average global temperatures rose by 0.74°C (1.33°F) last century, and that 14 of the 15 hottest years on record occurred during the 21st century.[6] If atmospheric carbon dioxide levels double from their pre-industrial levels as is likely, temperatures could rise by as much as 4.5°C (8.1°F) over the next century.[7] An even higher temperature increase has not been ruled out.[8]

At first thought, an increase in global temperature of 4.5°C (8.1°F) may not seem so serious. But small changes in average temperature can lead to significant climate effects.[9] At the poles, temperatures change more than two times faster than they do in the rest of the world.[10] During the last Ice Age, glaciers covered most of Canada and the northern U.S., but the earth was only 5°C (10°F) cooler on average than it is now![11] A warmer atmosphere magnifies hurricanes, typhoons, floods and wildfires.

The 2013 Fifth Assessment Report of the Working Group II of the IPCC, the world's scientific authority on climate change, has for the first time stated that it is quickly becoming less possible to avert catastrophic climate change by reducing new emissions than it was in 1997. The Fifth Assessment Report Summary for Policy Makers establishes that carbon remains in the atmosphere for hundreds of years, and removing the carbon that is already in the atmosphere is now needed to avoid catastrophic risks. Solutions that solely involve curbing current emissions are quickly becoming insufficient.[12] However, there is good news. For the first time in history, carbon negative technologies now exist, and are becoming commercialized.[13] Cleaning CO_2 from the atmosphere is now profitable, and it follows Adam Smith's own dictum about how new economic patterns are established. Adam Smith viewed individual profit as the driving force of our economy: cleaning CO_2 from the atmosphere can create profits and serve business interests. This is the foundation for believing that there is a solution for climate change.

Tipping Points and Potential Catastrophes

What climatic changes do scientists predict? The answer to this question is not straightforward. Although we are gaining more knowledge about climate change, there is still much we do not know. The current and projected rates of warming are terrifying because they are unprecedented over the last 20,000 years — in some cases millions of years. Climate is a dynamic and complex system that is difficult to predict. It is not simply the case that the earth will keep getting a little hotter each year and that each year the effects will be just a little worse than they were the year before. If this were true we could anticipate impacts and prepare for them. Instead, scientists warn, there are likely to be critical thresholds or tipping points in the climate system. These are truly points of no return; once we pass these thresholds, there is no turning back and the consequences could be dire. Complete collapse of the polar ice sheets or a change in ocean circulation, both of which are possible, would cause abrupt and catastrophic changes that no living or economic system could adapt to. If all of the ice sitting on land in Greenland and Antarctica melted, global sea levels would rise by 64–80 m (210–262 ft).[14] Although the complete melting of the ice caps is unlikely, even significant melting would put a multitude of highly populated coastal cities at risk of extreme flooding.

The Organization for Economic Co-operation and Development (OECD) recently ranked the largest cities at risk from sea level rises due to climate change; Miami and Shanghai were close to the top of the list with losses of $3.5 trillion and $1.7 trillion respectively.[15] The OECD damage estimates for cities alone include $35 trillion (in 2005 PPP USD) in property losses.[16] This is about 5% of current world economic output ($77 trillion) and yet only accounts for property damage — not loss of human life. The horrific 2017 hurricanes in the Caribbean and Florida are clear examples.

Scientists do not know exactly where the tipping points in the global weather system are because natural systems are highly sensitive to change and they are interrelated in complex ways. The earth's climate has never warmed this rapidly, by our own doing. We have no previous experience to draw on. We are taking a risk with the survival of our species; we are playing a game of Russian roulette with our climate system and the longer we play, the more likely it is that our luck will run out.

Researchers still know relatively little about feedback effects that might either increase or weaken the pace and effects of climate change. For example, the permafrost regions in Russia and Alaska contain high concentrations of methane, a GHG that, if released into the atmosphere, could accelerate the GHG effect. Western Siberia, which is one of the fastest warming places on earth, holds the world's largest peat bog, a region of permafrost the size of

France and Germany combined, which was formed at the end of the last Ice Age. It is melting, and scientists fear it could release as much as 70,000 tons of methane over the next several decades. Similar melting has already occurred in eastern Siberia.[17]

The earth is large, and there is an enormous momentum of hot air and water behind global warming. Because carbon pollution is so long-lived in the atmosphere, the emissions we are pumping now will continue to affect the earth for hundreds of years. The damage caused by human interference in the climate system may not be fully apparent for many years.[18] Warming by at least 2°C (3.6°F) by the end of the century is extremely likely.[19] We are already experiencing damage today, such as increased species extinction and the sinking of parts of Alaska under the warming seas.[20] Scientists have identified 12 diseases deadly to humans and wildlife that are expanding their geographical range due to climate change. These include avian influenza, Babesiosis, Cholera, Ebola, intestinal and external parasites, Lyme Disease, plague, Red Tides, Rift Valley Fever, Sleeping Sickness, Tuberculosis, and Yellow Fever.[21] Australia, the oldest of all the continents, experienced an epic drought from 2002–2009 that claimed many lives from wildfires and heat waves. The drought was followed by catastrophic floods, demonstrating that climate change does not cause a smooth transition to higher temperatures but rather amplifies violent fluctuations in weather conditions and climate.[22] According to Endangered Species International, the level of species extinction is now so high that we are in the midst of the 6th episode of species extinction in the entire history of planet earth. There have been five other episodes of this magnitude in the last 4.5 billion years, all of which were connected to climate change, but none of which were caused by humans.[23] This is the first major global extinction caused by humans.

Even if the global community acts fast to reduce carbon emissions, we may not be able to prevent catastrophic climate change by reducing emissions, since we procrastinated too long. For the first time, scientists are reaching a consensus that only removal of the carbon that is already in the atmosphere can work. In other words, limiting carbon emissions is needed but is no longer enough; we need carbon negative technologies, as explained in the IPCC Fifth Assessment Report.[24] In any case, to preserve the future we will have to reduce global emissions by roughly 60–70% of their current levels by 2050.[25] This is no easy task, but it is certainly worth doing. It is still possible. We may no longer be able to prevent climate change, but we can still prevent catastrophic climate change. Climate scientist Martin Parry warns, "we now have a choice between a future with a damaged world or a severely damaged world."[26] It is really just a question of degree. All of this can change, of course, if we use carbon negative technologies now.

Climate Change Impacts

Predictions of future climate change impacts range from significant and disruptive to potentially catastrophic. Many are surprised to learn that some of the most serious effects of the carbon pollution we spew into the air will be felt in the sea. Increased carbon uptake by the world's oceans has made them warmer and more acidic. This damages fragile marine ecosystems, especially corals and the many species that depend on them. About 50% of the coral in the Caribbean is already gone, and the same is true for a significant part of the Australian Great Barrier Reef.[27]

Warmer temperatures cause the seas to expand. Warmer temperatures also speed the melting of the polar ice caps. These two forces combined have already led sea levels to rise by 10–20 cm (4–8 in) from pre-industrial times, with recent estimates indicating that it may be at the upper end of this range. Because it takes so long for the oceans to cool, sea levels will continue to rise for centuries.[28] Global sea levels rose at a rate of about 1.8 mm per year from 1961–2003 and have accelerated over the past two decades. Since 1993, the rate has risen to about 3.2 mm per year. 2014 marks the highest sea level on record.[29] Complete melting of the Greenland and Antarctic ice sheets will displace millions of people. Under our current trajectory, Bangladesh and the Maldives will disappear and New York will sink beneath the waves. The Greenland ice sheet is now disappearing at more than 1.5 times the average rate that prevailed for the 20th century. Total collapse of the Greenland ice sheet is unlikely this century, but not impossible.[30]

When glaciers melt, they infuse the oceans with fresh water. This can damage fragile marine ecosystems and may disrupt ocean currents such as the Gulf Stream, which plays a crucial role in distributing heat on earth. Rising sea levels can contaminate underground fresh water supplies, a phenomenon that is already contributing to a shortage of drinking water worldwide. Rising sea levels have already contaminated underground water sources in many small island nations, as well as in Israel, Thailand, the Yangtze Delta in China, and the Mekong Delta in Vietnam.[31]

The impacts of climate change have also been felt on land. In Alaska and Russia, where air temperatures have risen twice as fast as global mean average temperatures, the permafrost layer of soil is melting, causing significant structural damage and sinking entire towns into the warming seas. In the mid to high latitudes of the northern hemisphere, snow cover has already decreased by 10%. The 20th century witnessed a two-week decline in the annual duration of lake and river ice-cover.[32] Glaciers, which contain most of the planet's freshwater supply, are retreating in many places throughout the world. Switzerland and South America,[33] for example, have currently lost between one-third and one-half of their glacier volume. Polar bears that depend on the ice to forage for food may soon become extinct.

A warming world may also witness major changes in precipitation, torrential rains, droughts, hurricanes, and typhoons. Arid and semi-arid regions can anticipate more severe droughts and desertification while other parts of the world can expect increased rainfall and flooding. The National Aeronautics and Space Administration (NASA) reported in August 2015 that the Central Valley of California is sinking due to drought by about 2 inches per month. This could prove disastrous for regions of the world already suffering from acute food shortages. The most dramatic declines in agricultural productivity are predicted for Africa and South Asia, where food shortages are most severe.[34] Early studies of climate impacts suggested substantial agricultural gains would occur in higher latitudes as a result of longer growing seasons and the fertilizing effects of higher carbon levels on crops. The results of more recent field studies, however, predict that overall crop yields will decline, even in regions where yields were once predicted to rise, such as the U.S. and China.[35]

Climate change is increasing both the frequency and severity of extreme weather events such as hurricanes, cyclones, heat waves and monsoons. Just as no single case of lung cancer can be blamed on smoking, no single storm can be blamed on climate change. However, many scientists interpret severe weather patterns in recent years as signs of climate change unfolding. The Rhine floods of 1996 and 1997, the Chinese floods of 1998, the East European floods of 1998 and 2004, the Mozambique and European floods of 2000, heat waves in western Europe in 2003, monsoon flooding in Bangladesh in 2004, hurricane Katrina in the

Atlantic in 2005, and Superstorm Sandy in New York in 2011 are sobering reminders of more powerful weather events to come as our planet continues to warm.[36]

Even the most conservative predictions of climate impacts suggest that most living systems will be adversely affected. Climate change poses indirect threats to human life by decreasing food and fresh water supplies and expanding the range of some vector-borne diseases, such as malaria and dengue fever. Catastrophic weather events, droughts and heat-related illnesses pose more direct threats. It is documented that 35,000 people perished in Europe during the heat wave of August 2003 — one of the hottest summers ever recorded in Europe. Last year was the warmest on record globally.[37] It is a sobering example of how vulnerable even the citizens of advanced modern economies are to unexpected climate changes. Fifteen thousand people died in France alone, where summer temperatures are generally mild and hospitals, retirement homes, and apartment buildings lack air conditioning.[38] Public health officials had no emergency protocols in place to deal with the crisis. Climate change has left public health and emergency officials worldwide scrambling to find the resources and the expertise to deal with climate-related disasters they never thought they would have to cope with. Sadly, this adaptive capacity is most lacking in the developing world, where the effects of climate change will be the severest.[39]

Climate change disrupts ecosystems and their abilities to provide critical ecosystem services, such as air, water, and food, which are essential for human health and well-being, and it will contribute to the wide-scale extinction of plants and animals.[40] We can already observe these effects. As northern climates become more like those of the south, some plant and animal species are shifting their ranges. Butterflies, dragonflies, moths, beetles, and other insects are now found in higher latitudes and altitudes where it was previously too cold for them to survive. Such changes can disrupt ecosystems in damaging ways. Across the western U.S. and Canada, climate change has led to unprecedented outbreaks in mountain pine beetles, resulting in widespread forest loss. The forest loss, in turn, reduces carbon uptake and increases emissions from the decay of dead trees.[41]

Rapid climate change has even led to genetic and behavioral changes in animal species. For example, in Canada the red squirrel now reproduces earlier in the year, and increasing numbers of blackcaps from central Europe spend the winter in Britain rather than the Mediterranean. But not all animal and plant species will be able to adapt. Species with longer lifecycles and smaller populations, such as polar bears, will experience a decline in their population and will go extinct as their land and food supply disappears with the changing climate.[42] It is difficult to predict how well living systems will adapt to changing climate patterns in the future. If warming is rapid and extensive enough, and circumstances and human settlements block their migration, plants and animals may not be able to adapt. Among the animals that may be most affected will be humans.

Can We Adapt to Climate Change? Mitigation and Adaptation

Unlike other living things that have always had to adapt to their environment in order to survive, humans have altered our environment throughout our history. Can we now adapt to the climatic changes we have unleashed? No one really knows the answer to this question. What we do know is that the more extensive the warming and the more rapid its onset, the less time there will be to adapt and the more dangerous the impacts will be.

We also know that adaptation must occur in a world that is already severely stressed. Population growth, deforestation, soil erosion, desertification, and wide scale extinction have diminished our resistance, leaving us more vulnerable to the effects of climate change. Our loss of natural storm barriers such as barrier islands, sand dunes, and mangrove forests, nature's first line of defense against tropical storms, increases the magnitude of property damage from storms. We saw this first hand with Hurricane Katrina in New Orleans, where almost all of the natural storm barriers had been destroyed.

Hundreds of millions of people live in areas that will be inundated when sea levels rise, such as the Nile Delta in Egypt, the Ganges–Brahmaputra Delta in Bangladesh, the Maldives, the Marshall Islands, and Tuvalu. Increased population density along the coasts is not just something we observe in the developing world. In the U.S., the migration to coastal cities and towns continues. More than half of the world's population now lives within 60 km (37 mi) of the sea.[43]

Billions of people worldwide already lack access to clean drinking water. Even industrialized countries expect fresh water shortages in the coming decades as water demand continues to outstrip available supplies. By 2025, two-thirds of the world's population could be living under "water stressed conditions."[44]

Climate change will intensify droughts, famine, and disease in areas of the world where millions of people — most of them children — are already engaged in a desperate struggle for survival. Asia and Africa have suffered from an increase in frequency and intensity of droughts in recent years. By 2030 it is predicted that almost half the world's population, including 75–250 million people in Africa, will be exposed to increased water stress — when the demand for water exceeds the available amount — due to climate change.[45] According to a 2013 study by the World Food Program, six million children, the majority of whom live in developing countries, die from malnutrition each year. In Bangladesh, one of the countries that will be hardest hit by climate change, 53,000 children already die each year from malnutrition.[46] Wheat production in India is being strained by climate change after years of robust growth. In some African countries, climate change is expected to reduce agricultural production by as much as 50% by 2020.[47]

Climate change will increase the spread of diseases such as malaria and dengue fever, which already claim millions of lives each year in developing countries. According to the DARA Climate Vulnerable Forum (www.thecvf.oegl), climate change causes 400,000 deaths on average each year, with Africa experiencing a disproportionate number.

Climate change will be most devastating to the world's poor and will kill more people in poor countries in Africa, South Asia, and Latin America than elsewhere. Extreme weather events are already more common in these areas. Population density in Asia, Africa, and Latin America is high, so when disaster strikes, more people feel it. Poor countries typically lack the resources — medical, rescue, and evacuation systems — to cope effectively with natural disasters.[48] Herein lies the cruel irony of the climate crisis. The countries least responsible for producing GHGs — most of the developing countries — are the countries now at greatest risk of death and human suffering because of climate change. The U.N. predicts that hundreds of millions of people in developing nations will face natural disasters, water shortages and hunger due to the effects of climate change.[49] Africa, which produces only 4% of global emissions, is the most vulnerable continent to climate change because of multiple stresses and its lack of capacity to adapt to change.[50]

The rich industrialized countries emit almost half of the global GHGs, even though they make up about 18% of the world's population.[51]

The second great irony of the climate crisis is that, for the first time in history, the welfare of rich nations will depend directly on decisions made in poor nations in Africa, Asia, and Latin America. Countries in these regions have the capacity to inflict, in the future, trillions of dollars in losses upon rich countries by industrializing and using their own fossil fuels, such as coal, as rich countries have done and continue to do. The challenge facing the global community is how to quickly reduce carbon emissions without leaving poor countries behind. The solution will demand nothing short of full cooperation between rich and poor countries. We have never achieved this in the past. Can we do it now? The Kyoto Protocol provided a common tool to unite rich and poor countries against climate change. What now? The Paris Agreement created a shared voluntary purpose. But President Trump withdrew the U.S. from the Paris Agreement.

Summary — What We Know

For too many years we have lulled ourselves into a false sense of security by treating the climate crisis as if it were still "uncertain" and of the distant future. It is true that the science is still relatively new and we cannot be completely sure about all aspects of climate change. We do not know for sure by how much the earth will warm, or how fast or how soon we will experience the impacts, or how bad those impacts will be and how well living systems can adapt. But we do now know more about climate change and we have greater consensus today than we do about most other scientific or social problems, such as dark matter in the universe or monetary policy. There is an alarming degree of consensus worldwide about the causes and effects of the climate crisis. In June 2015, Pope Francis wrote a 200 page encyclical *Laudato si'* about the reality of climate change, stating that it is caused by our actions and the world economy, and warning of the risks it poses to the survival of the human species.[52] Few things about climate change are uncertain at this point. And what is still uncertain may not become clearer to us until it is already too late.

Most importantly there is no longer much doubt that we are causing climate change. We cannot deny the devastating impacts our economic activities have on our planet and the crisis we have brought upon ourselves to solve. What we can do now is work cooperatively among nations to find solutions. And we must do it quickly. Scientists keep finding more evidence that warming is occurring faster, and that the effects are manifesting sooner than previously anticipated. As one eminent climate scientist warns, "we are all used to talking about these impacts coming in the lifetimes of our children and grandchildren. Now we know that it's us."[53]

If the governments of rich and poor nations cannot implement solutions because of political reasons, then it is up to individuals globally to start implementing their own solutions.

Here is a summary of what we know about climate change:

- The overwhelming majority of scientists believe that average temperatures on earth are rising because of humans and the globalization of industrial society.
- Burning fossil fuels is responsible for the observed increase in warming particularly since WWII, according to most scientists.
- The emissions we are pumping into the atmosphere now will affect the earth for hundreds of years.
- To avoid catastrophic risks, global carbon emissions must decline by roughly 80% by the year 2050. According to the International Energy Agency (IEA), this requires

an energy revolution involving \$43 trillion in investments in new power plants.

- The consequences of climate change range from disruptive to potentially catastrophic, with economic damages as great as 20% of global GDP at stake.[54]
- Climate change could cause \$35 trillion in damages (in 2005 PPP USD) to the largest cities in the world by 2070.[55]
- The ability of living systems to adapt to climate change will depend on the speed and severity of warming.
- Industrialized countries are most responsible for the risks.
- Poor countries are at the greatest risk of loss of life, property damage, and human suffering from climate change.
- Poor nations could inflict trillions of dollars in damages on industrial nations simply by using their own fossil fuels to grow their economies the way industrial nations do today.
- A total of 3,700 gigatons of CO_2 injected into the atmosphere will increase average temperatures by 2°C, and three times more in the North and the South Pole. Of these, 2,700 have been already injected, and there are at most 1,000 gigatons left in our carbon budget before we increase by 2°C.[56]
- We have already injected enough CO_2 into the atmosphere and it is sufficiently permanent that, at this stage, to avoid catastrophic damage we need to remove the carbon that is already there. In other words, carbon negative technologies are an imperative to resolving climate change.[57]
- It is possible to build power plants that are carbon negative, since the residual heat that power plants emit is enough to drive carbon capture from the atmosphere. The cost is sufficiently low that the captured CO_2 can be sold for profit in industrial gases markets, such as in the production of food and beverage, plastics, biofertilizers and other materials. It can also be used to enhance the productivity of oil wells.[58]

- There are some things we may never know about climate change until it is too late. Time is not on our side.
- If the ice sheets of Greenland and Antarctica melt, this will increase the sea level by 64–80 m (210–262 ft), according to the U.S. Geological Survey. This would flood much of the coastal area in the world and could displace hundreds of millions of people.

2

Insuring the Future

A crisis as serious as climate change demands immediate action. Why is it so difficult to convince countries to do more to prevent the worst outcomes when the science is compelling?

The reason is that the problem involves the use of energy — the key to economic production and the mother of all markets. Furthermore there is no simple solution to the problem. As Albert Einstein said, "problems can't be solved by the same mindset that created them." To solve the climate crisis we need to fundamentally rethink what it is that we produce and how we produce it. This basic truth is unsettling to many. However, it is not hard to imagine a world without fossil fuels. Wind farms, negative carbon, hydrogen cars, solar panels — the technologies of the post-carbon economy no longer sound as if they are out of a science fiction novel. The survival of the next generations is at stake. Our children understand that it is their world that we are changing. The future is now.

The Energy Dilemma

Energy is implicated in every life process. It is the single most important input in human production, especially in industrial societies. Until recently we have not had to consider the implications of our energy use. To solve climate change, however, we need to transform our energy use and our energy sources. We need to think about energy in new and innovative ways. We need to consider the impact of energy use on living systems today and in the future.

In pre-industrial societies human energy was the main energy source. Our physical capacity to do work was our energy constraint. The Industrial Revolution then gave rise to machines powered by wood, coal, and petroleum. Our use of these energy sources was constrained only by our ability to find them. For a long time the supply of fossil fuels seemed virtually unlimited. The world's population was small, roughly 800 million at the start of the industrial era, and new technologies were rapidly advancing that would improve our ability to discover and extract fossil fuels.[1] The consequences of burning fossil fuels and destroying the world's forests were mostly unknown to us. The scale of human production was small compared with the earth's capacity to absorb and recycle the waste we generated as by-products. The industrial society never considered the possibility of ecological limits to growth. It believed that nature was ours to conquer.

The Industrial Revolution enabled a dramatic rise in living standards and consumption in a period of rapid increase in population in poor nations, which was mostly driven by the introduction of antibiotics. By 1800, the world population was nearly one billion; by 1930, it had doubled. At the end of the 20th century the world population reached six billion,[2] and, according to the UN, we have now reached 7.35 billion. As the number of people on the planet grew exponentially, naturally so did the demand for energy, resources, and land. The scale of human production grew more than the earth's carrying capacity.

By the second half of the 20th century it became apparent that we were approaching certain ecological limits. Not only were we using up resources at a pace that was not sustainable, but we were generating too much waste. Whispers of an emerging environmental crisis could be heard. The signs were palpable: we were choking on smog in London and Los Angeles and many rivers in the U.S. were so heavily polluted with chemicals that they were actually catching on fire, such as the Charles River in Boston and the Cuyahoga River in Ohio. A new consciousness of the ecological impacts of human production was emerging, and out of that the modern environmental movement was born.

Climate change, however, was still a distant concern for most people. Scientific understanding of climate change, its causes, and its impacts was still tenuous. The toll that our fossil fuel dependence was taking on all living systems was not well understood. We were far more concerned with energy independence and diminishing energy supplies than we were with climate change. M. Hubbert's peak theory predicted that oil production would "peak" in the 1970s, at which point demand for oil would increase faster than the supply of oil worldwide.[3] The economic impacts of dwindling fossil fuel supplies loomed large in our minds.

Today, however, we understand that there is a problem far more serious and immediate than diminishing oil supplies. It has been said that the Stone Age did not end for lack of stones, and the same can be said about the era when fossil fuels reigned unchallenged. The situation is very different and uncertain now. Together with successful calls for divesting from fossil fuels in leading universities and funds around the world, the coal industry and coal driven power plants are moving towards economic extinction. It seems clear that we must transform our energy needs and energy sources. Before the last drop of oil is extracted from the ground, independently of when that happens, climate change may force us to abandon fossil fuels as our major energy source in order to save lives and property.[4] But getting there will be an enormous challenge.

Fossil fuels generate roughly 84% of the energy used in the world today.[5] They are the main source of carbon emissions. The most valuable of these fossil fuels is petroleum. Petroleum fuels our transportation systems and it is the basis for the plastic materials that are universally used today. Electricity itself is mostly generated by fossil fuels. Power plants that produce electricity from fossil fuels are the single largest source of emissions on the planet, representing almost 45% of all emissions in the U.S.

Economic development depends on the availability of energy sources and right now, this still means fossil fuels. Across history and across all nations of the world, there is a clear and direct connection between energy use and economic output. A country's industrial production can be measured by its use of energy. More growth means more energy use. Until now, any attempt to restrict carbon emissions involved reducing energy use. Until low cost alternatives to fossil fuels are developed, reducing emissions may mean reducing economic output.

Transforming the global economy will create new winners. There will be new ways to make profits that do not involve generating carbon emissions. Improvements in energy efficiency will save households and businesses money and will improve productivity. Smart investors already sense these opportunities. According to the United Nations and Bloomberg, the amount of investment capital flowing into renewable energy worldwide has seen impressive growth since the mid-2000s. Investment was just \$8 billion in 2005. In 2007, we hit \$150 billion and, since 2010, we have seen annual investments of over \$200 billion.[6] Even more recently, at the beginning of 2020, the investment strategies of the world's leading institutional investors, from Blackrock to Goldman Sachs to Bank of America, announced plans to make climate change central to their investment portfolios. Larry Fink, chairman and CEO of Blackrock, the world's largest asset manager, with $7.43 trillion under management, announced in his 2020 letter to CEOs that the climate crisis will reshape finance and that sustainability is now at the center of Blackrock investment strategy.[7]

Blackrock is not alone. Goldman Sachs plans to spend $750 billion and Bank of America has committed $300 billion to sustainable investments over the next decade.[8,9] Additionally, private-equity firms including KKR, Bain Capital, and Apollo Global Management are also featuring environmental, social, and governance (ESG),[10,11] which shows that, in 2019, investors put $20.6 billion into funds focused on environmental, social, and governance (ESG) issues, nearly quadruple the record of the year prior.[12] In fact, Bank of America has estimated that in the next two decades, there will be more than $20 trillion of asset growth in ESG funds, in which climate change investment is a major component.

Institutional investment firms are also not the only entities altering investment plans. In January 2020, the EU announced a plan to use $1.11 trillion for sustainable investments across the coming decade.[13] In its "European Green Deal" released in December, the EU laid out a roadmap to make the bloc's economy sustainable, to be climate neutral by the year 2050. As long as gains to some sectors offset losses to others, there will be no net loss to the global economy from our efforts to prevent climate change. The global economy may even be more robust, if the looming threat of financial disaster from the damages of climate change is removed.

The Solution Must Be Global

In February 2006, the British Prime Minister at the time, Tony Blair, succinctly summed up the problem of global warming, saying "no nation in the world would voluntarily agree to reduce its economic growth."

Blair is right: trying to reduce global carbon emissions is not an easy political or economic task. To significantly reduce emissions we will need to rethink how we power our factories, fuel our vehicles, and heat our homes. It may even require changes in our lifestyles — where we live, what we buy, and how we spend our leisure time. With appropriate planning and investment in energy efficiency and alternative energy sources, we can reduce our dependence on fossil fuels without compromising our quality of life. But this means tackling the climate crisis head-on, rather than scrambling after the fact. Which country will go first?

The truth is no country will voluntarily reduce its emissions, unless other nations agree to follow suit. This one fact, as crude as it may sound, is the single most important reason why we need all nations to sign up to a global agreement on climate change. Of course, there are compelling ethical reasons why nations should cooperate to solve climate change. As citizens of the earth, we all share the responsibility to preserve the earth's climate system upon which we all depend. However, ethical values may not convince countries to reduce emissions as much as they need to in order to prevent climate change; at least they certainly haven't thus far.

Why is it that the climate crisis demands a global response? There are many other environmental issues that capture international attention but do not require an international agreement to solve them. The destruction of the Amazon rainforest is a case in point. All nations have a strong interest in preserving the Amazon rainforest because of its unparalleled diversity of plant and animal species and its ability to produce oxygen, the lungs of the world. Only the countries that contain the Amazon rainforest within their borders, however, have any real power to prevent its destruction. If Brazil adopts measures that encourage deforestation, or turns a blind-eye to or is unable to stop illegal logging, there is little the international community can do. It cannot intervene in the resource management decisions of a sovereign nation, despite the implications of those decisions for the global community. Other countries can encourage forest conservation in Amazonia by applying political pressure, providing financial support or by discouraging the consumption of Amazon forest products by their own citizens, but ultimately there is little that most countries can do directly to stop deforestation in the Amazon.

Climate change is different. In this case, all nations are implicated since all nations produce carbon emissions. And carbon emissions transcend borders. Global warming is the result of cumulative global emissions. The atmosphere does not distinguish between emissions from the U.S. or China; the atmospheric concentration of carbon is the same all over the world. Carbon emissions have the same impact on the global climate no matter where they are produced. It also does not matter where we reduce carbon emissions. A ton of carbon reduced in the U.K. is as valuable for preventing climate change as a ton of carbon reduced in India.

Avoiding climate change is a perfect example of a global public good.[14] A *global public good* is something that once provided benefits all countries and no country can be excluded. Emissions reduction by any one country helps all countries to the extent that it may lower the risk of climate change. Countries pay the cost associated with their own emissions reduction but share the benefit with the rest of the world. What incentives do countries have to provide global public goods? Countries do not always behave altruistically.

No country alone produces enough emissions to cause global warming. But perhaps two countries could help prevent further damage. If China and the U.S., the world's largest greenhouse gas polluters, eliminated all of their emissions, global emissions would decrease by about 45%.[15] This would be a huge step in the right direction. Yet emissions reductions by other countries would still be needed.

This is why a global agreement to reduce carbon emissions is so critical. No country has an incentive to reduce emissions, since doing so will not prevent climate change unless other countries do the same. To make it worthwhile for a country to combat climate change, its actions need to be part of a larger, coordinated global effort that has the potential to stop global warming.

This is where carbon negative technology™ comes to the front stage, as it allows us to see CO_2 as an asset instead of a liability.

An international climate treaty gives countries the assurance that their climate combat efforts will not be in vain. An international climate treaty gives countries more "bang for their buck."

So why has it been so difficult to forge an international agreement to prevent climate change, when it is in every country's best interest to do so? Why hasn't the U.S. ratified the Kyoto Protocol? Because until now there had been no financial incentive.

Free Riding

We know that without a global agreement to control and capture emissions, climate change will continue unchecked.

The damages from climate change will be significant — potentially catastrophic — for all countries, even though some countries may be harmed more than others. We also know that no single country can prevent climate change. This is why no single country will voluntarily curb its emissions. Without a doubt, all countries benefit if an international agreement reduces the threat of climate change. All countries should prefer this outcome to its alternative: no treaty and dangerous climate change. But there is a third outcome that countries may prefer even more that we have not yet considered. It is called free riding.

Free riding is the option that lets a country think that it can have its cake and eat it too. Here's how it works: if all countries but one cooperate to avert climate change, the one country that does not cooperate benefits from the efforts of the rest of the world, without having to reduce its own emissions. It reaps all of the benefit without any of the sacrifice. Some might call this behavior despicable, but it is a rational response to the incentives as they exist. It makes sense for any one country to attempt to free ride, either by opting out of a climate treaty or negotiating within a treaty to do as little emissions reduction as possible. But if one country thinks it makes sense to free ride, all countries think it makes sense to free ride. And if all countries attempt to free ride, we cannot solve the climate crisis.

An international climate treaty such as the Kyoto Protocol helps overcome the tendency toward free riding by identifying each country's role in solving the climate crisis. But it cannot solve the problem entirely, since it cannot force countries to honor their commitments. We do not have international governance that is capable of this type of enforcement. When after years of negotiation and after signing the agreement in 1997, the U.S. walked away from the Kyoto Protocol process in 2001, there was little the international community could do in response.[16] To its credit the global community remains committed to solving the climate crisis, with or without U.S. participation.

Before we get too discouraged, let us remember that the global community has proven itself capable of negotiating solutions to major global environmental problems. The Kyoto Protocol is not the first international environmental agreement. Perhaps the best example of successful global cooperation to solve a transboundary pollution problem is the Montreal Protocol on Substances that Deplete the Ozone Layer. The Montreal Protocol phased out the use of ozone-depleting chemicals, thereby protecting the ozone layer from further deterioration. However, there were many factors working in favor of international cooperation in the case of the Montreal Protocol that are not present in the case of climate change. For one, there was more agreement and a better understanding of the causes of ozone depletion. Also, the consequences, especially the rise in skin cancers, posed an immediate and direct tangible threat to the inhabitants of rich industrialized countries. Secondly, from the perspective of industrialized countries, the benefits of protecting the ozone layer far exceeded the costs of eliminating ozone-depleting substances, even if they had to act alone without the help of other countries. It is critical that industry could find profitable alternatives to the chlorofluorocarbons (CFCs) that caused damage to the ozone layer. No such solution was found until now for fossil fuels, although new carbon negative technologies that remove CO_2 from air and sell it for profitable uses, such as food and beverages, polymers, and building materials, are a profitable answer that parallels what happened with the CFCs. Finally, the role of developing countries in the process was much less debated, since rich countries were the major producers of ozone-depleting substances.[17]

In contrast, negotiating the Kyoto Protocol involved more controversy and setbacks. And negotiations going forward will be anything but smooth. The Kyoto Protocol is a global landmark, perhaps the most important international agreement of our times. Together with the Montreal Protocol, these two global protocols created important precedents for resolving global environmental problems. The Kyoto Protocol gives us our best chance for solving the global climate crisis. The 2015 Paris Agreement indicates universal concern but has no mandatory tools to resolve the problem.

How to Price the Future: An Economic Perspective

There are many issues that lend themselves to a straightforward comparison of the costs and benefits of taking action. Unfortunately, climate change is not one of them. A cost–benefit comparison assumes that costs and benefits can be measured in monetary terms with a reasonable degree of confidence. The costs of emissions reduction, in principle, involve well-defined monetary expenditures that we can count. The benefits of preventing climate change, however, are generally more difficult to measure and to quantify. The benefits of preventing climate change are the avoided damages from climate change: lives and property saved. The problem is that many of these benefits are intrinsically priceless and unpredictable. It is a shortcoming of our economic systems that these benefits are difficult to compute.

Economists have long struggled with the problem of how to assign meaningful monetary values to seemingly priceless things such as human life, health, and ecosystems. Is a human life worth \$6 million or \$600 million? For a single person, their own life is priceless. Is a life saved in the U.S. more valuable than a life saved in India, because of the differences in expected lifetime earnings? Are polar bears worth more than Boyd's forest dragon, an Australian lizard expected to lose 90% of its habitat by 2050 from climate change, simply because more people recognize polar bears and think they are cute?[18] The answers to these questions may seem somewhat arbitrary — and to a certain extent they are. The problem of assigning meaningful monetary values to things that are not usually bought and sold in the marketplace plagues all cost–benefit analysis. This is a shortcoming that the carbon market overcomes for the care of the atmosphere. The problem, however, is particularly acute in the case of climate change, given the enormous potential for lives lost, plant and animal extinction and disruption to natural ecosystems.[19]

Discounting the Future

Even if we could all agree on a number that captures the benefit of saving lives, ecosystems and property from climate change, there is another fundamental problem to consider. Many of us will not witness the worst ramifications from climate change in our lifetimes. But our children might, and our grandchildren most certainly will, feel the impacts. The actions we take, or do not take, will have a much greater impact on those who are not yet born than on us. It is not just our own welfare at stake. Are we willing to pay today what it costs to ensure the welfare of future generations?

Almost everyone can identify a tendency to "discount" the future in decision making. We prioritize our present needs and desires and place less weight or significance on what may happen later on. The same logic applies in economics. It is common to treat a dollar of income earned today as more valuable than a dollar of income in the future. A bird in hand is worth two in the bush, is it not? We can invest our dollar today at a positive rate of interest and have more than a dollar in the future. We may be richer in the future, in which case a dollar will be worth less to us then than it is now. We may simply be impatient — we would rather consume with our dollar now rather than wait until later. These are all reasons why we may discount the benefit of saving future generations from climate change and place more emphasis on what it will cost us today to stop it. In neoclassical economics, there is an axiom that builds this in, and it is called "impatience." It was created by the great economist and Nobel Laureate, Tjalling Koopmans.

It is a standard practice in financial markets to discount the short-term future. A dollar paid next year is worth less than a dollar in hand today: this is why we pay interest on a bank loan. But the practice of discounting can paralyze us from taking action on global warming, which has long-term effects. Through discounting we greatly underestimate the costs we inflict and literally obliterate our concern for future generations. For example, if we cause $1 billion in global warming damages in the year 2011, these damages are worth $940 million in 2010 at a standard 6% discount rate and the same damages inflicted just one year later, in 2012, are only worth about $888 million in 2010. The value of these damages drop exponentially with time: $1 trillion in damages 100 years from today is a mere $3 billion today at a 6% discount rate. This is less than oil companies' earnings in one year and does not appear to merit much global attention. The problem is about the same no matter what discount rate we use: 6%, 5%, or even 2% or 1%. Using any fixed discount rate, discounting decreases exponentially the dollar value today of the costs we inflict on future generations. This is a problem of current economics.

Is it appropriate to discount the welfare of future generations? Economists have long struggled with this question. A well-known article on this subject was written in 1928 by Frank Ramsey, in which he argued, "It is assumed that we do not discount later enjoyments in comparison with earlier ones, a practice which is ethically indefensible and arises merely from the weakness of the imagination."[20] It may be appropriate to discount the future if we believe that future generations will be much wealthier than we are now and better able to cope with the effects of climate change. But what if they are not wealthier? Without taking action, we expect a planet severely diminished in its capacity to support human life and economic activity. In this case, the dollar we think is more valuable to us now could actually be worth much more to them. We should give them more, not less, since conditions may be so much worse because of climate change. For the first time in the history of industrialized nations, the next generation may not surpass the living standards of the current generation. Progress may be reversed. Human existence may be at stake.

But even if we were to assume that future generations will be as rich as we are, there is still little justification for highly discounting the future damages of climate change. Most of us would not consider our granddaughter to be less precious than our daughter, simply because she was born a generation later,[21] although we are currently behaving as if that was case. If we want to sustain our planet's ability to support future generations, we need to invest in climate change prevention now. Discounting may make sense for certain financial decisions but we cannot let the tendency to discount skew our perception of the urgency of reducing emissions in the present, not when the future health of the planet and the welfare of our children, and their children, is at stake.

The good news is that economic theory now exists that can tackle all the same economic problems without, however, discounting the future the way it has been done until now. In parallel with T. Koopmans' classic axioms, new axioms were introduced to define sustainable preferences.[22] It is possible now to replace discounting the future with an approach that treats the future equally with the past and the present. This provides new criteria that we call "sustainable preferences" to make choices about projects without assigning next year a weight that is a fixed smaller proportion of today's, which leads to exponential discounting.

Evaluating Risk

How do we evaluate the risks to future generations? Climate change presents risks that are poorly understood, dependent on our actions, collective, and irreversible. How can we determine what cost is worth incurring today to prevent these risks, when the risks we face from climate change are unknowable in advance?

Managing climate risk is not a new activity. Indeed, environmental uncertainty, such as weather risk, is the oldest form of economic uncertainty. In medieval England, a peasant farmer's land was broken into many widely-dispersed parcels. Historians interpret this as a way of hedging climate risk.[23] Land in different locations would be affected differently by droughts, floods, and frosts so by spreading land holdings over different locations, as well as by organizing agricultural cooperatives and buying insurance, farmers succeeded in managing climate risk.

However, today's concerns about global climate change break new ground in two ways.[24] One is that the scope of potential damages is global. Climate changes will affect large numbers of people in the same way. A rise in sea level, for example, will affect low-lying coastal communities across the globe. The second is that climate change risks are driven by human activity. Unlike the risks associated with earthquakes or volcanic eruptions, which are beyond our ability to control, the risks of climate change are dependent upon our actions,[25] how fast and how extensively we are able to curb our emissions. Climate has always been unpredictable but the inclusion of these two new elements has magnified the degree of uncertainty significantly.

The risks of climate change are essentially unknowable in advance. Climate change is a global experiment that cannot be repeated. With so much uncertainty about the expected damages from climate change, or alternatively, the anticipated benefits from climate change prevention, how can we estimate precisely how much it is worth spending to stop global warming? We cannot. Ranges of costs have been offered, but there is no precise estimate, no magic number that will somehow resolve any concerns about whether climate policy makes good economic sense. The decision to invest in climate change prevention is not one that can be framed in standard cost–benefit terms. There is too much at stake, too much that is unknowable, too much that matters that economics cannot appropriately account for and measure. Suffice it to say that if scientists are right about climate change and its consequences — and we have every reason to believe that they are — then it must make good economic sense to avoid it. The benefits of preserving the planet's climate system must outweigh the costs, even if we cannot measure the benefits precisely. How could they not?

Can We Afford the Future?

This is the trillion dollar question at the heart of the climate debate.

How much it will cost to prevent climate change is something we cannot really know in advance. The cost of preventing climate change will depend on how quickly we act; the longer we delay, the more we will have to reduce emissions in order to lower atmospheric carbon levels and the more costly it will be. If we are concerned about the costs of preventing climate change, stalling is definitely not the right strategy.

We also do not know for sure how quickly new technologies will emerge and what those new technologies will look like. Until the creation of Kyoto's carbon market, which is still in its infancy, there was little incentive to reduce emissions, since no one had to pay for them. The carbon market, explained in more depth in Chapter 4, rewards those who reduce emissions and punishes those who do not. This undoubtedly spawns new technologies and new ways of doing business. We are just not sure what they will be exactly.

We do have some estimates of what it may cost to prevent runaway climate change. These estimates should be taken with a grain of salt, but they at least give us some sense of what is at stake in the climate debate.[26] A recent report from the UN's Intergovernmental Panel on Climate Change estimates that stabilizing rising greenhouse gas levels would cost between 1–4% of global GDP by 2030.[27] This means that 1–4% of the value of what the world produces each year will have to be set aside to pay for emissions mitigation. In the scheme of things, this is not very much. But the costs are escalating as we speak. The longer we wait, the warmer the world will get, and the more we will have to spend to avoid the worst damages from climate change. In an influential report to the British government in 2006, former World Bank Chief Economist Sir Nicholas Stern estimated that that the world needs to spend 1% of global GDP each year to protect the future from climate change. Stern warned that if we fail to slash carbon emissions soon, the world will suffer losses ranging from 5–20% of global GDP each year. The longer we delay, the more likely that warming will surpass critical thresholds and the more devastating the losses will be.[28] For argument's sake, let us assume the highest cost scenario — that preventing climate change will cost as much as 4% of global GDP per year. Is it worth it? Where you stand on this issue may partly depend on where you sit. In rich industrialized countries, such as the U.S., the economy grows between 2% and 3% annually in a typical year. Losing 3% of GDP in any one year would mean that Americans would revert to the standard of living they had the previous year. This hardly signals a return to the Stone Ages. Had the U.S. spent 3% of its GDP on climate change in 2007, it would have cost $434 billion, or $1,440 per person.[29] No one will argue that these amounts are insignificant, but they are far less significant than they first appear. To help remedy the financial crisis, the U.S. approved spending more than $700 billion in its initial attempt to bail out cash-starved banks.[30] The war in Iraq cost the United States upwards of $2 trillion and Americans supported the war in Iraq much less than they do climate change mitigation; the approval for the war was approximately 32%, while 50% believe more should be done for climate mitigation.[31] The costs of protecting the future from climate damages may be well within the range of what most Americans are willing to pay.

In poor nations, losing 4% of GDP is a much harder pill to swallow. This explains the reluctance of most poor nations to agree to binding limits on their greenhouse gas emissions. The key to saving the Kyoto Protocol will be finding ways to shield poor countries from the costs of preventing climate change. We will explain more about how to do this in the chapters that follow.

Risks Too Great to Ignore

Now that we have some sense of what it may cost to prevent climate change, let us ask the question again. Can we afford the future?

The truth is we cannot afford not to. Whatever the exact costs of preventing climate change may be — 1% or 3% of global GDP — they pale in comparison to the potential costs of doing nothing. Failure to prevent climate change may cost the world as much as 20% of global GDP each year.[32] The economic impacts of climate change would be an extreme contraction on a level not seen since the Great Depression (the worldwide economic downturn originating in the U.S. in 1929). What a terrible legacy to leave our children and grandchildren. Spending 1–4% of global GDP to avoid damages as great as 5–20% of global GDP is clearly well worth it if one is realistic about the alternatives — which are non-existent.

Consider some more evidence. According to a recent study, if climate change continues unchecked, the U.S. will lose approximately \$270 billion, \$500 billion, \$960 billion, and \$1.87 trillion annually by the years 2025, 2050, 2075, and 2100 respectively from the combination of hurricane damages, residential property losses, increased energy costs, and water supply costs alone. By 2100, the losses associated with these four categories are estimated to account for 1.8% of U.S. GDP while the total cost from global warming could reach as high as 3.6%.[33] To put this number in context, the U.S. economy grows by less than 3.6% in most years. This means that the damages from climate change are significant enough to seriously lower living standards, and reduce economic growth.

Even more ominous is a recent study by the Organization for Economic Co-operation and Development (OECD). They estimated the economic damage that climate change could cause to major cities (see Table 1 "Top 20 Cities Threatened by Coastal Flooding from Climate Change") by the year 2070. They predicted \$3.5 trillion in property losses in Miami, \$2.1 trillion in New York City, \$1.9 trillion in Calcutta, and \$1.7 trillion in Shanghai. By 2070, 150 million people worldwide will be threatened by coastal flooding. The value of assets at risk to flooding alone could reach \$35 trillion.[34] The human cost of climate change will be greater still. The climate vulnerability monitor attributes more than 400,000 deaths and five million illnesses each year to climate change.[35] Alaska's towns are already

Table 1 Top 20 Cities Threatened by Coastal Flooding from Climate Change

Rank	Country	City	Potential Losses Today (2005 PPP USD billions)	Potential Losses 2070 (2005 PPP USD billions)
1	U.S.	Miami	416	3,513
2	China	Guangzhou	84	5,358
3	U.S.	New York–Newark	320	2,147
4	India	Calcutta	32	1,961
5	China	Shanghai	73	1,771
6	India	Mumbai	46	1,598
7	China	Tianjin	30	1,231
8	Japan	Tokyo	174	1,207
9	China	Hong Kong	36	1,164
10	Thailand	Bangkok	39	1,118
11	China	Ningbo	9	1,074
12	U.S.	New Orleans	234	1,013
13	Japan	Osaka–Kobe	216	969
14	Netherlands	Amsterdam	128	844
15	Netherlands	Rotterdam	115	826
16	Vietnam	Ho Chi Minh City	27	653
17	Japan	Nagoya	109	623
18	China	Qingdao	3	602
19	U.S.	Virginia Beach	85	582
20	Egypt	Alexandria	28	563

Note: Cities ranked in terms of value of assets exposed to coastal flooding in 2070.
Source: Hanson, 2010.

sinking under the sea as its permafrost soil is melting; the city of Newtok is being relocated at a cost of $380,000 per person.[36] The ravages of hurricanes Kartrina in New Orleans and Maria in Puerto Rico gave the world a sobering glimpse of the magnitude of human suffering climate change might bring. More than 1,800 people died in New

Orleans, while over a million were evacuated, and hundreds of thousands of people have still not been able to return to their homes.[37] The count in Puerto Rico is about four times higher.

Swiss Re, the world's largest reinsurer, confirms a general upward trend in the number, severity and costs of natural disasters. It reports that 12,700 people died worldwide from natural disasters in 2014 and 27,000 in 2013, most of them in Asia and most of them because of storms and flooding. Bangladesh and India accounted for 6,700 of those who perished. Property losses from catastrophes in 2014 topped $110 billion. Most of this damage was not covered by insurance, although insurance companies worldwide did pay $28 billion in damages from natural disasters in 2014 and 80% of insured losses worldwide in 2014 stemmed from natural disasters.[38] It is no wonder that insurance companies such as Swiss Re are calling on political leaders to find immediate solutions to the climate crisis.

Insurance for Future Generations

Reducing carbon emissions is like an insurance policy for future generations. It is a bet that we cannot lose. Even if climate change ends up causing much less damage than scientists believe, reducing emissions will have led us to invest in new industries, develop new jobs and new technologies, and increase efficiency and productivity. If it causes damage on the scale now predicted, we will have saved tens of millions of lives and trillions in property.

In the face of such risks, public policy must find preventive measures. It is like insuring a house against fire. We are not sure it will happen but it is prudent to insure against such a loss. People routinely insure themselves against risks with much lower probabilities of occurrence than climate change.

The chances of someone's house burning down or flooding are near zero; yet most people are willing to pay substantial amounts to insure their homes against these risks. Healthy young people who buy life insurance are another example. People routinely insure themselves against personal catastrophes that are much less likely than worst-case climate catastrophes for the planet. They are motivated by caution and a sense of responsibility to the loved ones they may leave behind. Their decisions account for the well-being of others alive today and others born in the future. Why shouldn't our decisions about climate change reflect the same prudence and commitment?

We should think of climate policy as an insurance policy against potentially catastrophic events. Our decisions should err on the side of caution. Compared with climate catastrophe, all other outcomes are irrelevant. And certainty regarding climate change impacts can only be achieved after it is too late.

If we treat emissions reduction as insurance for the planet and for future generations, the amount we need to spend to minimize the risks of catastrophic impacts makes perfect sense. The "insurance premium" in this case is the cost of decreasing the carbon emissions that industrialized countries are mostly responsible for.

Spending 1–4% of global GDP to reduce atmospheric carbon levels makes perfect sense from the insurance viewpoint. It is actually less than what the world currently pays to insure itself against catastrophes. Table 2 "Worldwide Insurance Coverage in 2013" details what the world paid in 2007, a typical year, in premiums for all non-life insurance policies. This includes insurance policies to cover losses from natural disasters such as floods, fires, and typhoons and man-made disasters such as plane crashes, rail disasters, and shipwrecks.[39]

Table 2 Worldwide Insurance Coverage in 2013

	Premiums ($ Millions)	Growth (%)	World Market Share (%)	Premiums as a % of GDP	Premiums Per Capita ($)
North America	1,391,106	−1.6	30.2	7.5	3,957.2
Latin America and Caribbean	178,022	5.7	3.9	3.0	290.9
Europe	1,619,997	1.4	35.2	6.6	1,822.7
Western Europe	1544,285	1.4	33.6	7.6	2,861.2
Central and Eastern Europe	75,712	0.8	1.7	2.0	235.1
Asia	1,252,376	−0.7	27.2	5.2	295.2
Japan and industrialized Asian economies	793,332	−4.3	17.2	10.9	3690.5
Emerging Asia	411,521	7.3	8.9	3.0	112.1
Middle East and Central Asia	47,524	6.2	1.0	1.5	139.9
Oceania	89,731	7.3	2.0	5.3	2,368.9
Africa	69,938	5.6	1.5	3.0	63.7
World	4,601,169	0.2	100.0	6.1	644.8
Industrialize countries	3,819,748	−0.9	83.0	8.1	3578.8
Emerging markets	781,421	6.1	17.0	2.7	127.9
OECD	3,744,960	−1.1	81.4	7.6	2,862.6
G7	2,881,263	−1.3	62.6	8.1	3,755.2
EU, 15 countries	1,435,714	1.3	31.2	7.9	3,251.5
North American Free Trade Agreement (NAFTA)	1,418,457	−1.5	30.8	7.1	2,992.2
Association of Southeast Asian Nations (ASEAN)	87,684	10.5	1.9	3.4	132.3

Note: This includes coverage for man-made and natural disasters but not life insurance.

Source: Swiss Re. Economic Research & Consulting, Sigma No. 3/2014.

Table 2 shows that, according to Swiss Re, the world already spends the equivalent of 6.1% of global GDP, $645 per person, on insurance premiums to protect against human-made and natural disasters. Is it surprising to learn that the world already spends this much on insurance against low-probability but costly disasters? It could be. For years we have listened to opponents of the Kyoto Protocol tell us that the costs of preventing climate change, which are well within this range, are beyond what countries can reasonably afford.

North America spends 7.5% of its GDP, $3,957 per person, on insurance policies to protect against risks that will be less damaging, on average, than climate change. Europe spends 6.8% of its GDP, $1,823 per person, on insurance. What are Europeans and North Americans protecting themselves from? Why, disasters, of course! Natural disasters, including floods, hurricanes, typhoons, and droughts which climate change will increase the frequency and severity of, accounted for 51% of all disasters in the world, 79% of all deaths from disasters and 82% of all insured losses in 2013.[40] Certainly no one in the U.S., Canada or Europe refutes the logic of spending this much on disaster insurance each year. Why shouldn't we apply the same logic to climate change?

In terms of a return on investment, the premium to avoid climate change is a better bet. The world spent 2.7% of global GDP, $2.0 trillion, on non-life insurance premiums in 2013.[41] Catastrophic damages that year amounted to around 0.2% of global GDP. And most of that damage was uninsured. Of the $135 billion in catastrophic damages, only $45 billion was insured.[42] In comparison, the premium for averting catastrophic climate change is 1–4% of global GDP. And at a minimum, we can avoid losses equivalent to 5% of global GDP now and forever, according to the 2006 Stern Report.[43] If the worst predictions about climate change impacts are true, our insurance premium helps us to avoid damages as costly as 20% of global GDP or more.

When individuals pay premiums for fire or flood insurance, they almost never see any of the money they paid come back to them. To insure against climate change, we pay today to avoid damages in the future. Our children and grandchildren will be the primary beneficiaries. But the premiums we pay to avoid future climate damages will benefit us in the here and now. Here's why. In the U.S. for example, the energy-related sectors of the economy are not particularly "labor intensive." This means that these sectors tend to employ more capital (machinery and equipment) than labor in production. Therefore money invested in the technologies and industries of the future produce more jobs and money-making opportunities than the entire oil and gas industry both in the U.S. and globally, based on the 2017 International Renewable Energy Agency. Investments in energy efficiency save households and businesses money on their energy bills and allow them to spend money on goods and services that can create more jobs compared with the jobs provided directly by the energy industry.

By all measures, insuring the future against catastrophic climate change is a prudent investment. The logic is so compelling there are only two possible reasons to deny it. Either we do not believe the science, or we discount the well-being of future generations. The science at this point seems compelling. Is it really that we care so little about the welfare of our grandchildren?

Who Pays for the Future?

Who should pay the premium for protecting future generations from climate change? As contentious as this question sounds, there is substantial agreement about the answer. The countries most responsible for the problem of climate change, the industrialized countries, should take the lead in global mitigation efforts. Not only are these countries to blame for about half of global emissions, but their higher incomes mean that they are better able to afford emissions reduction than countries in Africa, Asia, or Latin America where the majority of people survive on less than \$2 per day.[44]

The Kyoto Protocol is our insurance policy for the future and we need to perfect it, find appropriate ways to limit emissions globally and distribute most of the burden for paying for emissions reduction onto those who are most responsible. The Paris Agreement underscores this wisdom. Sound impossible? It really isn't once we understand the role the global carbon market can play in reducing emissions and distributing the costs.

The most important thing the global carbon market does is attach a price to carbon emissions. Until recently, there was no cost associated with producing carbon emissions; therefore, there was no incentive to reduce them. The price of carbon in the global carbon market is currently \$12 at market exchange rates and is expected to reach \$20 per ton of carbon in the 2020s. The world currently produces the equivalent of 34–38 gigatons of carbon dioxide per year. If we require emission producers to pay for every ton of emissions at a price of \$20 per ton, with 30 billion tons it will generate \$640 billion dollars each year.[45] *This is equivalent to roughly 0.8% of global GDP.* Does this number sound familiar? It is similar to the estimated cost of avoiding catastrophic climate change.

What does this mean? This means that using the carbon market to force emitters to pay for their emissions will generate enough income to offset the projected costs of preventing climate change. Remember how we agreed that it makes sense to pay a premium of at least 1% of global GDP to insure against climate damages? We now have a way to pay for it or most of it. The carbon market guarantees that the emitters will foot the bill and have an incentive to do so.

It also means that the global economy will be no worse off for finally taking the threat of climate change seriously. By attaching a price to something that was previously treated as free — carbon emissions — the carbon market creates a new income source that can pay for emissions reductions. Emission producers will be the ones to clean up their emissions. The carbon market allows us to shift the costs onto emission producers, while the net cost to them could be zero.

This can be a rather simple solution to a complex global problem.

Short Term Solutions — Long Term Challenges

Fossil fuels create a Gordian knot tying up three key global issues: energy security, economic development, and climate change. The fossil fuel age faces a cruel choice: economic development and energy independence clash against a stable climate. Today, we cannot have them all. The attendant geopolitical conflicts take several forms. Fossil fuels are the primary energy source in the world today. Because they are unevenly distributed on the earth's crust they have led to wars and conflicts, prompting understandable calls for energy security and independence. At the same time economic development still depends crucially on the use of energy, and in today's economy, this means fossil fuels.

In the longer term, the only way out is to disentangle the use of energy from carbon emissions, namely to make available clean and abundant renewable energy sources. But this is not feasible in the short term because of the sheer scale of the fossil fuel infrastructure: about $40 trillion today, and with current trends about $400 trillion by the end of the century.[46] The short term and the long term present different problems, however, and therefore require different solutions.

Time is not on our side. The Intergovernmental Panel on Climate Change (IPCC) scientists posit that atmospheric carbon concentration stabilization is needed, and that it will require a significant reduction in global greenhouse gas emissions by 2050.[47] Avoiding further carbon emissions in no way solves the short-term problem. Even if we stabilize at the current level of emissions, we still globally release carbon dioxide at a rate slightly above 32 billion metric tons per year and therefore will increase carbon concentration.[48]

The solution for much of this problem is negative carbon — a type of technology that is able to actually reduce carbon from the atmosphere in net terms. This is in contrast to technologies that simply reduce emissions, which at best leave the amount of carbon in the atmosphere unchanged. For instance, "clean coal," which is achieving a great deal of attention in the U.S. Congress and Senate, means coal that produces fewer or no emissions. The process of extracting that coal, however, is anything but clean.

Clean coal has at best a neutral "footprint" in terms of emissions that can leave atmospheric carbon unchanged. This may help as a stop-gap measure, if one forgets the other forms of environmental destruction that coal mining leaves in its wake. But even assuming this problem away for the moment, clean coal alone is not sufficient. Even if it was possible, it would not suffice to arrest catastrophic climate change. New coal plants that clean the carbon they emit are a step forward but they create burdensome economic costs and, in any case, they merely stabilize the implacable growth of carbon concentration at current rates. More to the point, such coal plants defeat the long-term objective of making an orderly transition to non-fossil resources. It is critical that short-term goals be compatible with long-term objectives. We must avoid the trap of defeating long-term aims by focusing solely on short-term targets. Capturing carbon dioxide directly from fossil fuel power plants may delay the time of reckoning but it adversely impacts the long-term objective of replacing fossil fuels with renewable sources and carbon removal.

The long-term solution we seek is to disassociate energy use from fossil fuels. This cuts the Gordian knot referred to earlier, which ties energy use, economic development, and climate change together. A long-term transition away from fossil fuels to alternative sources of energy[49] that are more broadly distributed can provide economic development and security without inducing global warming. A transition away from fossil fuel energy sources seems inevitable in the long term, because fossils are limited in supply. Alternative sources of energy are a necessary condition for sustainable development in the future and the rapidly growing world demand for energy will require a variety of alternative sources.[50] Supplies are not the problem. Through solar alone, the United States has the potential to supply more than 100 times the electricity it uses annually. Moreover, solar is a more democratically distributed input than other natural resources such as oil and coal.[51]

However optimistic one may be for the long term, it is important to recognize that this long-term solution is not appropriate for the short term. A transition to alternative energy sources is expected to take a long time since most of the energy used in the planet today is obtained from fossil fuels such as oil and coal.[52] As already pointed out, the change could take time and require a massive new infrastructure.[53] Yet as long as we continue to use fossil fuels and emit carbon we increase the concentration of greenhouse gases, and the risk of catastrophic climate change.[54]

What is the Short-Term Solution?

We cannot eliminate fossil fuels from our economy overnight. A quick and drastic reduction in emissions is not feasible due to the sheer size of the fossil infrastructure that needs to be replaced.[55] Indeed, rich and poor nations could be seriously affected by economic disruptions caused by a drastic decrease in the use of fossil fuels. Rapidly growing nations such as China and India are heavily dependent on coal; so are the U.S. and Russia. Hydroelectric power covers only 6% of world energy use, about the same as nuclear power, and renewable sources account for only 1% of the world's energy production today. It does not seem realistic to drastically decrease the use of fossil fuels in the short term, which is why there is an increasing call to capture the carbon emitted by fossil fuel plants and store it safely in the form of commercial products that create profits and employment.

In the long term, we must take into consideration that an alternative source should be able to provide 5 to 10 times the energy used in the world today. This is a standard projection of energy demand by the end of this century.[56] None of the five main types of renewable energy — hydroelectric, geothermal, solar, wind, and biomass resources — nor nuclear energy can offer this possibility, either because they lack the capacity or because to do so would create additional problems. For example, biomass for energy competes with food production, and is much less efficient per square meter than solar (about 3% of the energy potential provided by solar for the same surface area), and hydroelectric lacks the capacity and has environmental consequences. But solar energy — in particular Concentrated Solar Power (CSP) — could easily meet the demand with limited environmental impact. A combination of all of these energy sources that includes solar could therefore offer a reasonable long-term solution.

In the short run, according to the IPCC Fifth Assessment Report of 2014 and the 2015 Paris Agreement, we need negative carbon. This implies a way of reducing the atmospheric concentration of carbon altogether.

The technology strategy we need should accommodate both the short- and long-term goals, and the transition from the short into the long term. This is a tall order because such a technology must simultaneously facilitate the transition to alternative sources providing for massive increases in supplies for the long term, while in the short term allowing the continued use of fossil fuels and simultaneously decreasing the carbon content in the planet's atmosphere.

Among several available technologies, one called the Global Thermostat — created by Chichilnisky and Peter Eisenberger — has the capability to produce electricity while simultaneously decreasing carbon in the atmosphere by low cost air-extraction and storage (cogeneration of electricity and carbon capture).[57] In this process, the carbon concentration in the atmosphere decreases while producing electrical power. This patented (32 patents) process uses the residual process heat that remains after the production of electricity to capture carbon from the atmosphere. Electricity is produced usually by turbines driven with high heat (about 300°C (570°F)) and after the high heat is used, the residual low temperature (80°C) heat can be used to capture carbon from air. This process uses any source of process heat to cogenerate electricity and carbon capture (fossil fuels, nuclear or concentrated solar power plants, aluminum smelters, refineries, and others) and can make a fossil fuel power plant a "net carbon sink," namely a site that actually reduces atmospheric carbon.[58] Such a combination is unusual and contrasts with the physical realities of the fossil fuel economy today, where the more energy that is produced the more carbon dioxide is emitted. In contrast, the proposed technology reduces more carbon from the atmosphere, the more electricity power it produces. This provides real protection against human-induced climate change since it allows us to become carbon neutral in the short term and enables an orderly transition from the short term to a renewable energy future, enhancing energy security and economic development.

As we will see in the chapters that follow, the Kyoto Protocol ensures that developing countries are compensated for emissions reductions that take place within their borders. Rich countries can purchase certified carbon off-takes from developing countries through Kyoto's Clean Development Mechanism (CDM) and apply them towards their own emission targets. Negative carbon technologies could provide more financial compensation for developing nations through the CDM than simply stabilizing emissions. Global Thermostat plants would get credit both for the avoided carbon, from using a carbon neutral source of energy to produce electricity, and for the reduction in carbon dioxide that they provide through air capture and storage. Thus, the CDM can be a powerful tool in the financing of Global Thermostat Plants for developing nations. This in turn can provide developing nations in the long term with clean energy infrastructure, and in the short term it can provide a transfer of technology and a source of clean and abundant energy to grow their economies.[59]

Equally important, however, is that this type of technology can help level the playing field between poor and rich nations, while reducing the risks to all countries from climate change. The recent investment boom in poor countries resulting from the Kyoto Protocol's CDM has benefited some poor countries much more than others. Investments are now flowing into China to build hydroelectric, wind and, most recently, natural gas-fired power plants. Why China? Simple. This is where most of the developing world's emissions come from. Indeed, over 18% of world emissions come from China, while only 3% come from the entire African continent. This is natural in a nation that by itself represents 20% of the world population. But the CDM program was designed to fund changes to reduce emissions, and so 60% of all CDMs went to fund changes in China's energy structure, which produces large emissions, while leaving out the poorest nations in the world because they happen to emit so little. A similar situation emerged in India. This problem can and should be corrected by the use of carbon negative technologies, because even though a poor nation emits very little, with carbon negative technologies it can reduce CO_2 in the atmosphere more than what it emits, indeed much more than what it emits.[60]

Africa plays a lesser role in Kyoto's current CDM. It receives little today in the way of technology and wealth transfers under Kyoto because it consumes so little energy and generates too few emissions. Today, the Kyoto Protocol and the CDM are just about reducing emissions. And since so little reduction can be achieved in Latin America or in Africa there is little role for them to play.

But all this changes with negative carbon technologies. These could be located in Africa or in Latin America and could allow those regions to play a significant role in global climate change prevention efforts. With negative carbon, Africa could significantly reduce carbon in the atmosphere, becoming an excellent candidate for CDM projects (perhaps even better than China). Will this happen? Will Africa be able to capture 30% of the atmosphere's carbon dioxide even though it emits only 3%? Can Africa save the world? To answer this, we must first explore the Kyoto Protocol, its carbon market and its CDM. If developing nations are offered funding from the CDM to clean the atmosphere — to remove more carbon than they emit — they are likely to promise to limit their emissions to what they can achieve with the funding and the technology available to them. A willingness by developing nations to agree to this new form of mandatory emissions limits would in turn help overcome the main hurdle created by the U.S. Congress in the unanimously passed Byrd–Hagel Resolution. The U.S. could now accept mandatory emissions limits in a way that is consistent with the limitations established by the Byrd–Hagel Resolution, which requires that the U.S. accept no mandatory limits unless the developing nations do. The new carbon negative technology can overcome this obstacle.

3

The Kyoto Protocol and its Carbon Market

The Kyoto Protocol creates incentives for new technology and has the potential to change the way we use energy and resolve global warming. It breaks new ground. It is important to remember that it is the first international agreement based on the creation of a new global market, a market for trading rights to use the planet's atmosphere.

The carbon market became international law in 2005 and can become the largest commodity market in the world. According to Bart Chilton, commissioner of the U.S. Commodities Futures Trading Commission, "even with conservative assumptions, this could be a $2 trillion futures market in relatively short order."[1] Chilton believes that the carbon market could trade futures, and its trading volume could reach $2 trillion dollars in the near future. The carbon market has some unique properties. It is a market based on the trade of a global public good — the reduction of global carbon concentration in the atmosphere. The properties that distinguish global public goods from private goods that are commonly traded, such as grain, houses, machines and stocks, have important implications for market behavior. In a market for global public goods, equity and efficiency are inextricably tied in ways that can unite the interests of rich and poor, businesses and environmentalists.[2]

Carbon markets are controversial institutions. It has already been said that many businesses fear them but in reality they are quite simple. How do they work?

Each trader has emissions limits; those who over-emit have to buy rights from those who under-emit. This penalizes the bad guys and compensates the good guys with minimal government intervention. In a nutshell, the idea is to use Adam Smith's famous invisible hand,[3] the hand of the market. This invisible hand can join the interests of the business sector with the social interests of environmentalists like nothing has done thus far. A similar but different system has worked successfully at the Chicago Board of Trade since 1993. Its sulfur dioxide (SO_2) market decreased acid rain in the U.S. in a very cost-effective way, using a simple cap-and-trade system that is quite different from the carbon market. This is explained in Chapter 4.

Yes, make no mistake. The carbon market is all about profits, which Adam Smith identified as the driving force of the capitalistic economy. This is Adam Smith's green hand, profits that clean the atmosphere. Adam Smith's hand needs a bit of help to do its magic. The market could not function without binding emissions limits to the traders. These are property rights.

The purpose of caps on industrial nations' emissions is to guarantee that global emissions will not exceed levels that could cause catastrophic climate change — the so called "carbon budget." Industrial nations emit about half of carbon emissions in the world today, and the Kyoto Protocol provides caps for all industrial nations. This is the feature of the carbon market that environmentalists favor. Of course the Kyoto cap requires that the nations that signed the Protocol in 1997, which include the U.S., ratify the Protocol and obey the emissions limit accordingly. Almost all industrial nations have ratified the Protocol, including Australia, which held out until 2007 and changed positions several times. More recently Canada dropped its participation.[4] When the Protocol was created in 1997, 160 nations voted for it under the condition that it would have to be ratified by nations representing at least 55% of global emissions. The U.S. was the exception, which voted for the Protocol in 1997 but has not yet ratified the Protocol. The U.S. agreed at the U.N. meetings in Bali in December 2007 to join the Kyoto Process and arrive at a solution by the end of 2009. Neither of these events happened. President Barack Obama's administration announced his intention to make this matter a priority and to use the cap-and-trade system. Under his administration the U.S. Supreme Court ratified the Environmental Protection Agency's (EPA) right to mandate carbon emissions limits, and to impose such limits for stationery sources, like power plants, in 2013 and 2014. These EPA limits intend to cut carbon dioxide emissions from existing coal plants by as much as 30% by 2030.[5] Those limits can be the basis for a Federal Carbon Market in the U.S. The U.S. is the second largest emitter in the world and historically the first emitter, and therefore its ratification of the Kyoto Protocol or its support for carbon limits and the carbon market would have a measurable impact on global emissions. In 2016, the Supreme Court stayed EPA regulation until several states resolve legal challenges, an unprecedented move.

Nevertheless, great progress has been achieved in recognizing the problem and the value of the carbon market in resolving it. In 2015, CEOs from the six largest gas and oil companies (BG Group, BP, Eni, Royal Dutch Shell, Statoil, and Total) wrote openly to the United Nations encouraging them to "introduce carbon pricing systems where they do not yet exist" and "creat[e] an international framework that could eventually connect national systems."[6] In June 2015, Pope Francis produced an encyclical *Laudato si'* identifying climate change as caused by humans.[7] But for many business leaders, carbon markets are still a source of fear and loathing. The fears are about sharp increases in the costs of doing business, especially for electricity and commodity producers, which are central to the economy but generate substantial carbon emissions. These businesses will have to pay, to some extent, for their emissions. There are also concerns about the volatility of carbon prices when the market starts trading. Carbon prices would increase some commodity prices and business costs. The concerns of private industry can be real and must be addressed.

At the same time, climate change is a real concern as is the continued massive extraction of resources from the world's fragile ecosystems. Some business leaders anticipate potential gains in technological innovation and profitable opportunities arising from the carbon market. They are right. The largest investors in the world are betting a substantial percentage of their risk capital in the renewable energy sector, a sector that has increased in importance from an average of roughly 3% of Silicon Valley venture capital investments in 2002–2005 to over 15% since the end of the financial crisis.[8] Clean energy is arguably the fastest growing area of business in the world today and already provides more jobs in the U.S. and globally than the entire oil and gas sector, as validated by the 2017 International Renewable Energy Agency. The level of investment in clean energy reached $270 billion in 2014, and some predict that by 2030, the level of these investments will reach as high as $630 billion,[9] replacing existing power plant infrastructure valued at over $45 trillion.[10]

We have seen that some within the business community are in favor and some are against the carbon market and its price for carbon. How can business opinions be so polarized? Is the carbon market a villain or a hero? Both views are right. Each looks at the carbon market through a different lens: the "before" lens and the "after" lens. Before the carbon market, the uncertainty about the carbon price can be damaging to business. There are perceived costs and no benefits. Yet it is only after the carbon market starts to operate that a price signal can be created. A price signal refers to a market price that signals through its level (high or low) the real costs, the real scarcity or the real value of a commodity or whatever is traded. For example, before the carbon market is created, there is no market price to signal the real cost of emitting carbon. There is no signal about this cost because there is no market price for carbon. Once the carbon market is introduced and trading begins in the European Union Emissions Trading Scheme (EU ETS), a market price emerges (ideally, about $30–60 per ton of carbon emitted) and the cost of carbon is "signaled" through the carbon price to the entire economy.[11] This carbon price signals how costly it is to the economy to emit carbon. The price signal makes cleaner technologies more profitable than the rest: clean technologies do not pay for emissions, emitters do. Therefore, when the carbon market operates, Adam Smith's green hand rewards under-emitters and penalizes over-emitters, aligning business and environmental interests.

The opportunities are open to all, and the rewards can be very substantial once the transition from the "before" to the "after" is completed. The International Energy Agency (IEA) says an energy revolution is required, arising from worldwide restructuring of the power plant sector, and that restructuring the $45 trillion power infrastructure will become a major business opportunity.[12] Using the economic incentives created by the carbon market to "make a profit while doing good" would give a major boost to businesses. Business could react very positively to such a market-based scheme.

Unless there is a global agreement on mandatory emissions limits, however, there is little reason for a single nation to limit its emissions, as no nation can resolve global warming on its own. And therefore there is little reason for any nation to have a national carbon market on its own. In addition, national carbon prices will always match global carbon prices, so no national market can set prices on its own. All regional carbon markets will derive support from the Kyoto Protocol global carbon market, and none will exist as efficiently without it. It is as simple as that.

The global carbon market became international law and started operating for industrial nations when the Kyoto Protocol came into force in 2005. This is also called "the U.N. carbon market"; it traded $176 billion per year by 2011 but has gone down in recent years as the Kyoto Protocol emissions limits for the Organization for Economic Co-operation and Development (OECD) nations have come to their projected end. Yet the market retains a high potential for growth.[13] This market is the crucial component of the Kyoto Protocol that sets it apart from all other international agreements. And it is the force compelling Australia, the U.K. and soon possibly the U.S. to introduce their own national carbon markets. One of the largest U.S. states, California, already has an official carbon market. In 2015, China announced it would adopt the carbon market and several regional carbon markets were created and confirmed in 2017. Together with the EU Emissions Trading System (EU ETS), this means that the largest segment of carbon emissions in the world is now under carbon market regulation.

How Kyoto's Carbon Market Works

Because of its historic importance, the negotiation of the Protocol naturally involved drama, suspense, and intrigue. However, as Raul Estrada Oyuela the Protocol's negotiator, reminds us, the Kyoto Protocol is the product of 30 months of complex negotiations and of a climactic last minute adoption.[14] Its articles and paragraphs, therefore, need careful interpretation and further elaboration.

Step one: Limiting emissions, nation-by-nation

The Kyoto Protocol is based on the principle that international cooperation is needed to combat climate change. This was set forth by the 1992 United Nations Framework Convention on Climate Change (UNFCCC), which explained the objective of the Kyoto Protocol, as follows:

> *The ultimate objective of this Convention and any related legal instruments that the Conference of the Parties may adopt is to achieve ... stabilization of greenhouse gas concentrations in the atmosphere at a level that would prevent dangerous anthropogenic interference with the climate system. Such a level should be achieved within a time frame sufficient to allow ecosystems to adapt naturally to climate change, to ensure that food production is not threatened and to enable economic development to proceed in a sustainable manner.*[15]

The first step for negotiators of the Kyoto Protocol was therefore to establish quantified emissions reduction commitments from participating nations. The total level had to be enough to reduce the threat of catastrophic climate change. Countries had agreed to voluntary emissions reductions at the 1992 Earth Summit but most nations failed to meet their voluntary targets, and emissions in most countries actually increased.[16]

By the time the Kyoto Protocol was negotiated and signed in 1997, it was very clear that voluntary targets did not suffice and binding limits were necessary.

Without a global cap on emissions, there can be no global carbon market. The carbon market trades "rights" to emit carbon into the atmosphere. These rights establish who has the right to emit what. Trading cannot begin until there is a clear agreement on the number of tons of carbon dioxide that each nation has the right to produce. Each seller must be able to demonstrate it has "title" to the carbon emissions rights it sells. This means that every nation must have a well-determined limit, otherwise it could sell infinite amounts and no market would exist.

Therefore, the Kyoto Protocol specifies the amount of carbon that each country can emit. These are written as a required percentage reduction from 1990 emission levels (see Table 1 "Annex I Emissions Target") for Annex I countries, which are industrialized countries and countries with economies in transition. They had to reduce emissions by an average of 5.2% of 1990 levels during the period 2008–2012.

Table 1 Annex I Emissions Target*

Country	**Target (in %)**
EU-15**, Bulgaria, Czech Republic, Estonia, Latvia, Liechtenstein, Lithuania, Monaco, Romania, Slovakia, Slovenia, Switzerland	–8
United States***	–7
Canada, Hungary, Japan, Poland	–6
Croatia	–5
New Zealand, Russian Federation, Ukraine	0
Norway	+1
Australia	+8
Iceland	+10

Note: * Target is defined as percentage change from 1990 emissions level during the commitment period, 2008–2012. ** The 15 member states of the EU have reached their own agreement about how to distribute the 8% emissions target amongst themselves. ***The U.S. has announced its intention to not ratify the Kyoto Protocol.

Some nations, such as Australia, could initially increase their emissions from 1990 levels.

At first glance, a reduction of 5.2% below the 1990 levels sounds like a very modest target. But compared with the 24% increase in global emissions expected to be the "business as usual" scenario by 2009, the real reduction required by the Kyoto Protocol was actually closer to 30%.[17]

The Kyoto Protocol counts net emissions. Each nation gets credit for carbon removals by natural sinks (sinks are reservoirs of carbon that reduce atmospheric concentration of carbon, such as oceans and forests) — from land-use activities and forestry. It states:

> *Net changes in greenhouse gas emissions ... resulting from direct human-induced land-use change and forestry activities, limited to afforestation, reforestation and deforestation since 1990 ... shall be used to meet the commitments under this Article of each Party included in Annex I.*[18]

This provided a degree of flexibility in reducing emissions, and much needed incentives to adopt sustainable land-use practices and conservation of forests. Annex I countries are effectively penalized if they deforest, since emissions from deforestation will count against their Kyoto emissions targets.[19]

As negotiators soon realized, it became difficult to establish a global emissions cap that is both fair and capable of preventing climate change. Global emissions caps were not easy to achieve. They had to meet scientific targets and protect poor nations. The Kyoto Protocol had to include policies and measures in such a way as to minimize "adverse effects" on climate change and trade, as well as social, environmental, and economic "impacts" on other parties, especially developing nations.[20]

The carbon concentration in the planet's atmosphere has a distinguishing and unusual feature: it is the same throughout the world. In other words, there is no way that one nation could choose one carbon concentration and another nation a different one. Whether they like it or not, all nations are exposed to the same carbon concentration in the atmosphere. It does not matter how much a nation emits or what they can afford. The atmosphere does not distinguish between emissions produced, or emissions reduced, in the U.S., China, Bolivia, or Australia. Each nation emits a different amount, this is true, but the ultimate carbon concentration that we are exposed to is the same for everyone on the planet. The laws of physics — nature's laws — reign supreme and trump the geopolitical and economic realities of energy use.

As has been said, this unusual feature of carbon dioxide makes the carbon market unique: it makes it a market for a global public good. The quality of the atmosphere is the ultimate equalizer and unifier; it is one and the same for everyone on the planet.[21]

In setting a global emissions cap, poor nations may be forced to accept a lower level of global emissions than they can afford. From this unusual feature it follows that developing nations must be treated more favorably than industrial nations in terms of what they emit.[22] Unless we all agree, the world will never reach the lower level of CO_2 in the atmosphere that we need in order to prevent climate change. This is why developing nations initially had preferential rights to emit. Forcing developing nations to emit less proves to be inefficient for the entire world as well as unfair, unless offsetting measures are implemented. This feature also makes the carbon market unlike any other market we ever saw before; it is a market where efficiency and fairness are tied to each other.

Carbon markets or carbon taxes?

Why did the Kyoto Protocol embrace the market approach to emissions reduction? The Protocol could have simply assigned emissions rights nation-by-nation to cap global emissions at acceptable levels. It could have stopped there, without creating an institution in which those rights could be traded. Indeed, in 1997, the debate in Kyoto was heading in that direction until the American negotiators insisted that there would be no agreement without emissions trading. The U.S. demanded additional flexibility and this is what the carbon market provided.

Think of it this way: the world needs to limit emissions no matter what. Each nation has to agree to limit its emissions; this is a must. Otherwise we cannot make a dent in climate change.

But once emissions limits are established nation-by-nation, the market approach becomes a very natural add-on. It has little cost and great benefits. Trading the right to emit allows a flexibility that is not available otherwise. A nation could one year be above its limit and the next below it — it is difficult to predict. With the market approach, as the total of all nations' emissions is below the global limit, each nation can fluctuate in its emissions — up one year and down the next. This is a natural and desirable flexibility. It certainly aligns well with the interests of the U.S., one of the two largest emitters in the world, for whom flexibility is valuable. This simple rationale won the day in Kyoto in December 1997, when the U.S. signed the Protocol together with 160 nations.

The foundations for the carbon market began with the assignment of binding emissions limits to participating countries. This established a firm cap on global Annex I emissions — something that most environmentalists favored. This alone makes the market approach more attractive than carbon taxes, since taxes cannot guarantee aggregate emission levels.[23] This is a well-known and universal truth.

Indeed, one of the main differences between the carbon market approach and emission taxes is the degree of assurance they offer about world pollution levels. With a carbon market system the aggregate level of pollution is fixed by the total number of emissions rights issued. For example, if global emissions rights are capped at 6 billion tons of carbon dioxide, and if the system is enforced, total global emissions will not exceed 6 billion tons. That is all. The international community will always know how much emissions reduction will be achieved in advance. The total amount of global carbon emissions is predictable. But there is also an important aspect of the market approach that cannot be known in advance: the marginal cost to emitters of reducing their emissions to the specified level. This cost is reflected in the price of the emissions permit. This price is determined by the forces of supply and demand. In general, the price in the carbon market cannot be predicted in advance with any accuracy.

Contrast this to the situation with a carbon tax. The cost to the polluter is given by the tax and is known with certainty. But the aggregate amount of pollution cannot be predicted. As long as the cost of reducing a ton of carbon is less than the tax an emitter has to pay to emit that ton, the emitter will reduce emissions. Since we do not know in advance what the cost of reducing emissions will be, we have no way of predicting how emitters will respond to the pollution tax and by how much they will reduce their emissions. This is a key difference between a carbon market approach and a pollution tax.

Here is a simple example: consider a tax on cigarettes. A cigarette tax does not guarantee a reduction in smoking, it just provides a disincentive to purchase cigarettes. In theory, if we raise the cost of smoking people will reduce their consumption of cigarettes. This is the hope. But if we fail to set the price high enough, or if people choose to smoke despite the penalty, the consumption of cigarettes will not decrease. It could create a black market for cigarettes. It could then even increase consumption. It is a similar story with income tax and with estate taxes. In the case of income taxes, we cannot predict that people will work less, and in the case of estate taxes, we cannot predict that people will leave less inheritance because of estate taxes — and even if they did, we cannot predict by how much.

The same is true with carbon taxes. Carbon taxes will penalize emissions producers and provide incentives to reduce emissions, with the aim that carbon emissions subside. But carbon taxes give us no guarantee in advance that global emissions will decrease as much as we need them to do in order to avoid the risk of climate catastrophe. Given the urgency of the climate crisis and what is at stake, we do not have the time or leeway for error to keep adjusting the carbon tax until we settle on the tax rate per ton of carbon that induces our desired emissions reduction. This could take 30 years. Regarding climate change, one thing we know with reasonable confidence is how much global emissions must decrease by to minimize the most serious risks of climate change. It makes sense to start with what we know and build our strategy around it. This is what the carbon market allows us to do.

In situations of great environmental sensitivity to pollution, knowing the aggregate level of pollution that will result from a policy may be essential: this is an argument for emissions trading. The latter point is crucial to understanding why the carbon market, rather than the carbon tax, is more appropriate for addressing climate change. There are critical thresholds in the planet's climate system. If we pass these thresholds the consequences are irreversible. We need the assurance of a fixed global emissions cap to minimize the risk that we will surpass these critical thresholds.[24]

Another powerful political sensitivity surrounds the creation of a global taxing authority, which would be necessary should the carbon taxes approach be followed. This seems almost impossible to visualize, let alone achieve. It could be as difficult as the creation of a second United Nations. It is difficult enough for citizens in nations such as the U.S. to accept a global authority on international security issues; it is even more difficult to accept a global authority that would tax the U.S. A global tax authority does not exist and is never likely to exist. It would be universally opposed. The concern is the creation of a global bureaucracy that collects funds from all the nations in the world, corresponding to about 1% of the world gross domestic product (GDP) or about $1 trillion, and allocates them appropriately to avert global warming. How likely is it that such a tax bureaucracy will emerge any time soon? Lack of trust in a global governmental entity of this sort could sink the entire effort. By contrast, the carbon market sails easily through these difficulties because by its own nature it needs no bureaucratic intermediaries. The bad guys who over-emit pay the good guys who under-emit, simply and directly. There are no tax authorities in the middle, collecting funds and deciding what to do with them. There only has to be an authority that enforces the agreed mandatory emissions limits, but we need to stick to the mandatory emissions limits anyway to prevent climate change.

More generally, it is the politics and the physical realities of climate change that mitigate in favor of a carbon market, rather than a carbon tax. The U.S. found in the Kyoto negotiation of 1997 that the market-based approach is consistent with its prevailing market-orientated approach to economic policy. Tax-based approaches are anathema to a political climate in Washington that is strongly predisposed against taxes. Without the inclusion of the carbon market in the Kyoto Protocol, the U.S. would have walked away from the table. In Europe, the tradition is quite different. Carbon taxes are more consistent with the European approach to economic policy. Most European governments historically have had no natural affinity for market-based approaches to pollution management, perceiving markets to be part of the problem rather than part of the solution. Hence, the concept of a cap-and-trade regime was much less familiar in Europe. This, in part, explains the hesitancy European negotiators expressed about Kyoto's carbon market.

Step two: Allocating emissions rights

In establishing a global cap on emissions, the negotiators of the Kyoto Protocol also had to decide how to distribute those rights between countries. One would think that the two issues are separate, that finding a global cap is a different issue from deciding who gets to emit what. Surprisingly, however, this is not so. It turns out that these two issues are actually closely linked physically, economically, and politically. These links are quite important for understanding the challenges and the opportunities presented by the global climate negotiations.

In practice, the two issues were closely linked in the negotiation of the Kyoto Protocol and these links permitted the Protocol to be successfully negotiated in the first place. In December 1997, global caps were negotiated at the same time as each nation's accepted cap, nation by nation. It was extremely important for reasons of equity to provide preferential treatment to developing nations, for otherwise they would not have agreed to the lower emissions that the Kyoto Protocol was all about. Developing nations feel it is a matter of historical fairness that they should not be asked to clean up the carbon emissions that industrial nations produced during their own period of industrialization. Historically, the rich OECD nations, which house less than 20% of the world's population, emitted the majority of the world's emissions because they consumed the majority of the world's energy[25]; currently they still emit about 45% of the world's carbon emissions. The inequity in the world's use of resources is already a source of friction. Asking the developing nations to clean up after the industrial nations would only add to the conflict. There is an additional practical argument. Even if developing nations were to stop all their carbon emissions this would still not make a dent in the global warming problem because the majority of developing nations emit so little. In fact, all African nations together emit 3% of world emissions, and a similar amount is emitted in Latin America.[26]

For many people, this information may induce cognitive dissonance, because two popular and erroneous myths still prevail:

(1) Poor countries have larger populations and therefore, consume most of the world's energy.
(2) GDP per capita and carbon emissions are unrelated.

However natural they may seem, these two arguments are completely untrue. To address the first issue: according to the IEA, as of 2003, developing nations consumed about 60% of the world's energy. This might seem high, but keep in mind that the developing world contains over 80% of the world's population. The industrial nations with less than 20% of the world's population use about 45% of the world's energy.

Rich nations are responsible historically for the large majority of the world's carbon dioxide emissions. The rich nations' overconsumption of natural resources manifests itself in most other areas as well: metals, minerals, forest products. According to the OECD database, OECD nations consume approximately one-third of the world's meat.[27] And meat production and transportation cause about 18–25% of the world's carbon emissions, more than the entire transportation sector of the world economy including all cars, ships, airplanes, and trucks.[28]

Emissions rights are highly valuable commodities, especially when those rights can be traded in a global carbon market. The allocation of emissions rights is a potent tool for redistributing wealth between nations. The more emissions rights we give a country, the more fossil energy it can use and the more income it can earn by selling this valuable commodity in international markets. As we will see in Chapter 5, we can harness this potential to close the global income divide between the poor and rich nations.

Who should reduce carbon emissions: the rich or the poor?

Whereas the size of the global emissions cap should, at least theoretically, be based on climate science, the issue of how to distribute the emissions rights has an interesting feature: it can be used to ensure efficient market solutions.

This provides an objective answer to a seemingly subjective question about how to divide emissions rights between countries. The division of rights is associated with the total amount of rights for the world as a whole. In the early part of the 20th century the great Swedish economist Eric Lindahl explained why efficiency dictates that lower income people should have more rights to use public goods.

British economists Arthur Pigou and James Mirrlees explained this issue as well in the context of taxes. In 1992, Chichilnisky, Heal, and Starrett came to a similar conclusion about carbon markets.[29] In simple words, when dealing with public goods, equity is connected with efficiency. Perhaps not coincidentally, the negotiators of the Kyoto Protocol came to exactly the same conclusion in December 1997.

Negotiators could have adopted several different approaches to allocating emissions rights. For example, one approach would be to establish caps based on equal percentage reductions for all nations. In that case, a country like Bangladesh, where the average citizen makes $1,100 per year, would have to reduce its emissions by the same percentage as a country like Germany, where per capita income is $47,600.[30] Given the vast income disparities that persist worldwide, this approach is considered highly unfair by most.

Others have proposed allocating emissions caps on the basis of population size. In this case, China, with a population of 1.37 billion, would be given many more rights to use the atmosphere than the U.S., with a population of 320 million. They would be given four times as many rights under this scheme. Many have argued that this is fair, but it may not be politically feasible. It is hard to imagine the rich nations of the world ceding that much advantage to the more populous developing nations. The right to emit carbon is, at present, the right to use fossil fuel energy, such as coal, of which China has some of the largest deposits in the world.

In practice, how did negotiators establish emissions caps and allocate emissions rights between countries in the crucial days leading to the December 11, 1997 Kyoto Agreement? Once again, negotiators were guided by the principles of cooperation established by the UNFCCC, which happen to agree with Lindhal and Pigou, as well as in general terms with the work of Chichilnisky, Heal, and Starett.[31]

The UNFCCC established the principle of "common but differentiated responsibilities," which recognized that nations have contributed unequally to the build-up of atmospheric greenhouse gas (GHG) concentrations in the past, and that nations have different abilities to pay for emissions reductions in the present. This principle is the source of the tension between rich and poor countries over whether and how to cap developing countries' emissions after Kyoto emissions limits expire in 2020. Article 3 of the UNFCCC states:

> *The Parties should protect the climate system for the benefit of present and future generations of humankind, on the basis of equity and in accordance with their common but differentiated responsibilities and respective capabilities. Accordingly, the developed country Parties should take the lead in combating climate change and the adverse effects thereof.*

Article 4 of the 1992 United Nations Framework Convention on Climate Change (UNFCCC) establishes that developing nations will not be required to reduce their carbon emissions unless they are compensated for it.

Accordingly, the Kyoto Protocol did not assign mandatory emissions limits to developing nations. This was in recognition of their limited energy use historically and their special needs and limitations. It assigned emissions limits to industrialized countries, the 39 Annex I countries that together account for two-thirds of global emissions. Within the group of industrialized nations, individual country commitments vary; but on average countries were required to reduce emissions by 5.2% of 1990 levels during the first commitment period, which ran from 2008 to 2012.

The Kyoto Protocol instructed the industrialized nations to lead the global effort to combat climate change. It demanded that they go first, blaze the path, and share their acquired knowledge and technologies with developing nations so that they too can follow in the near future. But let us remember that the Kyoto Protocol is the product of international negotiation. The different roles it assigns to developed and developing nations reflect the will of the international community. Kyoto is truly remarkable. Not only did it cap global GHG emissions for the first time in history, but it also convinced nations to lay aside differences and embrace a strategy that distributed the burden of emissions reduction fairly but unequally for practical reasons. It made concessions to equity that are almost unprecedented in international affairs. These trailblazing results set precedents that could change the world economy in the 21st century.

Step three: Flexibility and efficiency

To design an international climate agreement that enough countries could support, the negotiators had to solve two issues simultaneously: (i) reduce global emissions to avert climate change and distribute them fairly, and (ii) achieve the greatest degree of flexibility and secure efficiency to achieve the targets at the lowest possible cost. The carbon market solved both issues at once. In fact, the Kyoto Protocol reduced CO_2 emissions of participating nations by 22.6%, when comparing 2012 emission levels to 1990 levels.[32] This did not happen for the nations outside of the Kyoto Protocol, which rapidly increased their emissions.

Around $176 billion was traded annually by the carbon market in its peak in 2011,[33] and, in total, $130 billion was transferred through the Clean Development Mechanism (CDM) to developing nations.[34]

Just how does the carbon market guarantee flexibility and efficiency? Even though it is distinct from all other markets, the carbon market is still a market. By their own nature, markets can be efficient mechanisms. This is why markets are the most powerful institutions in the world economy today.

The carbon market has the potential to allocate resources efficiently in the economy. By placing a price on carbon emissions, the carbon market forces us to come to grips with an underlying scarcity we previously had ignored, the atmosphere's finite capacity to absorb our GHGs. It makes us confront the real costs of using fossil fuels, costs that include the damage we inflict on ourselves and on future generations.

It may seem distasteful to assign property rights and a price to such a precious resource as the planet's atmosphere, but there is no real alternative. The world economy is a market economy; it relies on the price signal to determine the "best" or "most efficient" use of scarce resources. Effectively, the price of carbon in the global economy had been zero until 2005. It has been said that "markets have never got a price as wrong as they have the price of carbon!" As it has also been said, "A cynic is someone who knows the price of everything and the value of nothing."[35]

Since carbon emissions are a by-product of our energy use and energy is the most important input for economic production, the failure to price carbon correctly throughout history has led to the inefficient use of resources until now. It is the reason we are in the predicament we are today, racing against time to wean ourselves off from fossil fuels. Fossil fuels are very costly to humankind; the costs include climate change; if we include human lives lost, species extinction, and irreparable damage to ecosystems, the costs are incalculable. Yet our markets so far have been ignorant of these costs. Imagine where we would be today if we had paid for carbon throughout our history. It is difficult to imagine how different our economies and technologies would look now.

The CDM and Developing Countries

The logic of emissions trading is actually quite simple. Suppose that you have two similar countries, Ecoland and Greentopia, and both are required to reduce their emissions. Ecoland could eliminate all the emissions originating within its borders, and Greentopia could do the same to all emissions within its borders. But what if the costs of reducing emissions in Ecoland are lower than the costs of reducing emissions in Greentopia? We know that there are some technologies for reducing emissions that will cost less than others. Technologies that improve energy efficiency, such as solar and wind energy, fluorescent light bulbs, improvements to the energy efficiency of homes and buildings, and fuel-efficient vehicle standards can conserve fossil fuel use at relatively low cost. Sometimes these options are referred to as "low-hanging fruit." What if low-hanging fruit still hangs in Ecoland, but Greentopia has already exhausted its low cost abatement options? What if, compared with Greentopia, Ecoland has abundant hydroelectric or wind power capacity that it can substitute for coal or natural gas? There are many possible reasons why one country may be able to curb emissions at lower cost than others. In which case, why not save money by having more emissions reduction take place in the countries where it costs least?

Emissions trading makes it possible to shift some of Greentopia's emissions reduction to Ecoland, so that both countries benefit. Either Greentopia can invest directly in emissions reduction inside of Ecoland and apply the emissions reduction to the total amount it is required to do under the Kyoto Protocol, or Ecoland can reduce its emissions and sell its unused user rights to the atmosphere to Greentopia. In either case, Greentopia meets its emissions target at a lower total cost, and Ecoland benefits by selling the emissions rights it does not need because it has either the resources, the technology or the untapped energy efficiency potential to reduce emissions at relatively low cost.

The Kyoto Protocol allows for emissions trading between industrialized countries — those that have binding emissions caps. The Kyoto Protocol also created a mechanism known as Joint Implementation (JI). This is a bilateral form of emissions trading between industrial countries (especially the transitioning economy countries in Central and Eastern Europe) that is project-based, allowing one country to invest directly in emissions reduction activities in another. However, because JI is bilateral trading, and it is unsightly to pair up a powerful rich nation with a poor developing nation in bilateral trading since the power relation seems too skewed, it applies only to industrialized countries — those countries with binding emissions limits. Indeed, Article 6 of the Kyoto Protocol states:

> *For the purpose of meeting its commitments under Article 3, any Party included in Annex I may transfer to, or acquire from, any other such Party emission reduction units resulting from projects aimed at reducing anthropogenic emissions by sources or enhancing anthropogenic removals by sinks of greenhouse gases in any sector of the economy...*

To provide developing countries with incentives to reduce emissions, and to encourage investment and move technologies from the global North (developed countries) to the global South (developing countries), as the Kyoto Protocol was required to do, a mechanism for including developing nations in the global carbon market had to be designed.

The CDM, often referred to as the "Kyoto Surprise," provides the crucial link between developed and developing nations in the global carbon market.[36]

The CDM enables an industrialized country to invest in reducing emissions in developing nations and count the emissions reduction towards its own emissions cap. It is similar to JI, but it is not bilateral trading as it goes through the carbon market, which is a multilateral market. It defines a role for developing countries in the emerging global carbon market, and it allows for the profitable participation of the private sector. As defined by Article 12:

> *The purpose of the clean development mechanism shall be to assist Parties not included in Annex I [developing nations] in achieving sustainable development and in contributing to the ultimate objective of the Convention, and to assist Parties included in Annex I in achieving compliance with their quantified emission limitation and reduction commitment...*

The idea for the CDM is straightforward enough. Emitters can save money by investing in lower cost emissions reduction options in developing nations. In return, developing countries benefit from the direct investment in their economies and technology transfers.

In practice, there are several ways that emissions reductions can be financed through the CDM. Developed countries may finance project activities in developing countries and use the resulting emissions credits towards their own emissions targets. Emissions producers within developed countries — for example, utility companies — may invest directly in CDM project activities in developing countries and use the resulting emissions credits to demonstrate compliance with their own country's emissions restrictions. (In order for a country to meet its Kyoto target, it requires emissions reductions by emissions producers within its borders.) Alternatively, developing countries themselves can invest in their own emissions reduction projects and market the resulting credits through the CDM. Lastly, CDM project activities may be financed by third parties — often non-governmental organizations (NGOs), development agencies or private for-profit entities — and the resulting emissions credits can be sold in the global carbon market. To date, there have been CDM projects involving reforestation, hydropower, methane capture, energy efficiency improvements, and fuel-switching. Examples of such CDM projects are in Chapter 6.

The CDM connects global emissions reduction to the broader goal of sustainable development. It creates incentives for developing nations to adopt clean technologies and move into the future on a more ecologically sustainable development path than the one the industrialized nations used during their development. Yet the CDM remains one of the most controversial parts of the Kyoto Protocol. Why?

The main issue is that so far just under 50% of the CDM projects have gone to China.[37] This is because the projects that qualify for CDM credits are those that reduce carbon emissions. Since China is by far the largest emitter of all developing nations (approximately between 25–30% of the world's carbon emissions), it has the most emissions to reduce. Africa as a whole emits only 3% of the global emissions and therefore, today, there is little in terms of CDM that Africa can capture. It can reduce little in terms of emissions because it emits so little. The same is true in Latin America. It is one thing to reduce emissions, and something completely different to remove carbon from the atmosphere. As described in Chapters 2 and 5, what is required here is to introduce CDM projects that allow "negative carbon," namely those projects that can reduce carbon over and above what the region emits as a whole. In that case, with carbon negative technology, Africa could reduce 20% of world emissions even if it emits only 3% itself. This is the solution, but it requires a modest modification of the CDM to allow for such technologies. In 2009, at COP15 in Copenhagen, such a modification was introduced.

Avoiding the Problem?

A main criticism of the CDM is that while the EU must adapt to cleaner forms of energy, the CDM projects are sheltering the EU from going through this technological change. Another important point is that it is imperative that emissions credits resulting from CDM investments compensate the nation for "reductions in emissions that are additional to any that would occur in the absence of certified projects."[38] During the debate in Kyoto, the terms "paper tons" and "hot air" were used to indicate the importance of verifying emissions reductions in all countries participating in trading — especially developing countries. The Kyoto Protocol itself requires verification of the reductions by the appropriate CDM Accreditation Committee that resides in Bonn. The reason is simple. The CDM allows industrialized countries to substitute emissions reduction within their borders for emissions reduction in the developing world. If the emissions reduction credit they have purchased in the developing world is not based on honest-to-goodness emissions reduction, global emissions will not decrease. If the emissions reduction credits are not legitimate, for example, if countries are selling emissions credits for activities that do not reduce emissions, or for activities that do reduce emissions but which would have taken place anyway without another country's investment, then the CDM could undermine Kyoto's global emissions cap. As with all things where there is money to be made, the incentives for businesses to exaggerate on their claims of deserving carbon credits are high. This is not an indictment of the logic behind the CDM, rather a cautionary tale about its practice. Effective monitoring is, and will continue to be, essential to the judicious use of the CDM, but there is no reason to believe that effective monitoring is not possible.

There is a misunderstanding regarding emission limits in developing nations. While it is true that developing nations do not have emissions limits in the 1997 Kyoto Protocol, they do have *emission baselines*, which measure expected trends in business-as-usual emissions against what the CDM projects the reductions to be. The baselines are then used by the CDM Accreditation Committee to measure the extent of emissions that would occur without the project.

There is another issue to consider. There are concerns in the EU that the Kyoto Protocol was designed to shift the burden of emissions reduction more heavily onto the countries that are most responsible for climate change and that are best able to pay for emissions reduction. Will emissions trading allow industrialized countries to escape the burden of costly emissions reduction and forestall the transition away from fossil fuels? Prior to withdrawing from the Kyoto Protocol, the U.S. had indicated that it would purchase most of its required emissions reductions abroad. Although countries can use emissions trading to avoid investing in emissions reductions at home, fortunately, there is a solution: because emissions trading makes it easier and less expensive for countries to reduce and meet their emissions caps, we can lower the caps further in the next round of negotiations. We know further emissions reductions will be needed after Kyoto limits expire. Because carbon emissions trading lowers compliance costs, it increases the likelihood that countries will agree to lower caps in the future.

Nevertheless, the carbon market and its CDM remain controversial, as is shown in the next chapter. The carbon market is a new type of market, the likes of which have never been seen before in the world, and we are learning by doing.[39] It is not surprising that there is much to adjust to; Kyoto is, after all, only a first step in a long road.

Ratification and Implementation

Before the Kyoto Protocol could enter into force and become international law — before the emissions limits could become binding — the Kyoto Protocol had to be ratified by a number of nations. This had to include enough industrial nations to account for at least 55% of the emissions of that group. This provision was necessary to ensure that emissions reduction would be sufficient, in total, to make it worthwhile for countries to participate. In practice, this provision did not give any individual country the ability to veto the Kyoto Protocol, but it did indirectly give veto rights to the U.S. and Russia together. The Kyoto Protocol could become international law, or enter into force, without Russia or without the U.S., but not without both.

The U.S. is still holding fort against Kyoto, but Russia changed its position. It took eight years from its creation in 1997 for the Kyoto Protocol to become binding. It took this long for enough countries to ratify the Kyoto Protocol. The EU, despite its hesitancy over the inclusion of the carbon market, ratified the Kyoto Protocol only months after its negotiation. The U.S., one of the early signers of the Kyoto Protocol and the chief impetus behind its carbon market, announced its withdrawal from the Kyoto Protocol in 2001, under the leadership of the newly elected Republican president George W. Bush. The U.S.'s decision dealt a serious blow to the Protocol. Most economists in the U.S. deemed Kyoto dead. It took a great deal of conviction to believe that Bush and his eight-year presidency of the most powerful nation in the world were not as powerful as an international agreement that could stop climate change. Kyoto was humble but it has been so far more powerful than Bush. President Bush has now left office while Kyoto is international law. The carbon market is now present in some form in four continents, it is law in California, and it was most recently adopted by China. It was worth just over $176 billion dollars in the EU ETS in 2011,[40] the dollar amount that the market traded that year.

The ratification of Kyoto shifted all attention to Russia, the only other country with sufficient emissions in the base year to qualify the Kyoto Protocol for implementation. Russia was in a unique position. De-industrialization following the collapse of the Soviet Union in 1991 had resulted in a significant emissions reduction. This meant that Russia would have an easier time meeting its Kyoto emissions target than it had anticipated. It also meant that Russia was in a position to sell its unused emissions rights to other countries through the carbon market.

Russia ratified the treaty in late 2004. The long-awaited Kyoto Protocol entered into force 90 days later on February 16, 2005. As of early 2009, 181 countries have ratified the Kyoto Protocol, including 37 Annex I countries,[41] representing 64% of the emissions of industrialized countries. Australia ratified Kyoto in 2007 and since then, Canada and Australia dropped it. As already mentioned, notably absent is the world's second largest greenhouse gas producer, the U.S. The U.S. had signed the Kyoto Protocol in 1997. As of this writing, the U.S. is the only Annex I country that has not yet ratified at any point the Kyoto Protocol. However, there has been progress; the U.S. has an official carbon market in California and several eastern states, and since 2014 a law has restricted emissions from stationary sources such as power plants. That is progress. In 2017, however, the new Trump administration quickly returned to the past, withdrawing from the Paris Agreement, a move to be finalized in four years, by 2021.

The State of the Carbon Market

The global carbon market became international law when the Kyoto Protocol entered into force in 2005. How well has the global carbon market performed since then? Does its reality match its potential? A few basic numbers summarize how well the carbon market is performing thus far: the carbon market grew in six years from zero to $176 billion in annual trading in 2011.[43] It began to fall, however, as the Kyoto Protocol mandatory emissions limits approached their deadlines. Currently, approximately 40 national and over 20 subnational jurisdictions, including China, are putting a price on carbon. This represents between 40% and 50% of annual global GHG emissions under the jurisdiction of various forms of carbon markets.[44]

There are several regional carbon markets, all of which were made possible by Kyoto's provisions for emissions trading and the CDM. The EU ETS is the largest carbon market at present although China's will dwarf it in due course. Sales and re-sales of EU emissions allowances among nations in this market reached $25 billion in 2006 and grew to about $176 billion in 2011. California, the most populated state in the U.S., has an official mandatory carbon market. There are also several on the East Coast of the U.S. As already mentioned, China has adopted regional carbon markets. The Chicago Climate Exchange (CCX) and the New South Wales market (NSW), smaller markets for voluntary reductions by corporations and individuals, have witnessed record trading values and volumes, among regional carbon markets, since their creation. They grew strongly to an estimated $100 million but they are not expected to play a defining role in the global market.[45] Voluntary emissions limits have been proven not to work.

With respect to poor nations, the real success story was, and continues to be, the trade in emissions credits from project-based activities through the CDM, and to a much lesser extent, JI. China continues to dominate the CDM market, accounting for about 60% of the number of projects. The

World Bank reported that CDM projects were worth $130 billion in 2012. In total, over $130 billion has been transferred to developing nations by the CDM for clean productive projects that reduce the equivalent of 12% of global annual emissions.[46] This is a record transfer.

Who is buying and selling in the carbon markets? The main buyers in the carbon market are:

(1) European private buyers interested in EU ETS.
(2) Government buyers interested in Kyoto compliance.
(3) Japanese companies with voluntary commitments under the Keidanren Voluntary Action Plan.
(4) U.S. multinationals operating in Japan and Europe and preparing in advance for the Regional Greenhouse Gas Initiative (RGGI) in the northeastern U.S. and the mid-Atlantic or for the California Assembly Bill 32, which aims to establish a state-wide cap on emissions.
(5) Power retailers and large consumers regulated by the NSW market.
(6) North American companies with voluntary but legally binding compliance objectives in the CCX.

In 2006, European buyers dominated the CDM and JI markets. They accounted for 86% of the market, an increase from 50% in the year 2005. Japanese purchases were only 7% of the CDM market. The U.K. led the market with about 50% of project-based volumes, followed by Italy with 10%. Private sector buyers, predominantly banks and carbon funds, continued to buy large numbers of CDM assets, while public sector buyers continued to dominate JI purchases.

In addition to financial performance, it is important to evaluate the physical impact of the carbon market in order to keep track of the actual carbon reductions that the market achieves. The CDM market played a main role in this reduction. In contrast to the highly volatile 2006 EU ETS market, project-based credits demonstrated greater price stability, while the volume of transactions grew steadily. But most important is the fact that since 2002, a cumulative 920 million tons of carbon, an amount equivalent to 20% of EU emissions in 2004, have been reduced through CDM projects in developing nations at a value of $130 billion.[47] These trends continued and strengthened in 2007, with an additional $15 billion in CDM projects, although once again the majority went to China and very little to Africa or Latin America. Since 2015, the carbon market and its CDM softened, mostly due to the ending of the mandatory limits by 2020.

Carbon Prices and Market Stability

After a period of extreme volatility, the price of carbon is now relatively stable but low due to the lack of mandatory emissions limits beyond 2015. Since 2015, the price of carbon has hovered around 5–7 British pounds consistently in the EU ETS.[48] This was a main source of concern for private industry, which seeks firm price signals in order to plan for costs and opportunities. Because the carbon market is so new there is understandably some confusion about how carbon prices are set. Many believe erroneously that they are set by free floating supply and demand. While it is true that prices do fluctuate somewhat in the short term with supply and demand forces, it is possible to identify market "fundamentals" that determine carbon prices.

The key to explaining fluctuations is given by the behavior of the carbon markets in 2006, where a drop in carbon prices from $30 per ton to $10 per ton followed the selection of higher emissions caps by the EU. It is possible to show how carbon markets function to determine carbon prices, and how these prices fluctuate over time.

The main point is that, in a fossil-fuel dominated global economy, there are two "fundamentals" that determine prices in the carbon market: (i) emissions caps, which are a measure of scarcity and the extent of demand for "permits to emit" and (ii) the technology that allows us to transform fossil fuels into goods and services, which provides the "opportunity cost" of reducing carbon.

Let us consider caps first. The caps are provided by governments, in accordance with their Kyoto Protocol obligations. Governments can decide their own pace for achieving their Kyoto targets and they can establish their own caps accordingly. The lower the caps, the higher the obligation to reduce carbon, and therefore the higher the price of carbon. This is how the market operates. In 2006, the EU discovered that carbon prices were dropping because the caps on carbon emissions were set too high. The EU promised to adjust these caps correspondingly. By lowering emissions caps, the governments of the carbon market nations increased the demand for permits and increased the price of carbon. The lack of the mandatory limits after 2020 is weakening the carbon market, and the 2015 Paris Agreement, which is voluntary, has not helped.

The second determinant of carbon prices is the technology that transforms energy into goods. This provides the "opportunity cost" of reducing emissions, namely the goods that we fail to produce because we use less fossil energy, or the cost of carbon capture as discussed in Chapters 2 and 5.

This is how the carbon market works: it provides incentives for the use of clean technologies. It favors technologies that emit less carbon over and above those that emit more. The latter has to pay for its emissions, the former gets rewarded for avoiding emissions.

There is a critical interplay between carbon markets and technology. Technology has an impact on market prices. In a competitive market, market prices reflect marginal costs. Reciprocally, the carbon price has an impact on what technology is developed and used. This interplay is at the core of our ability to resolve the global warming issue in the short and the long term, because by using the right technologies we can avert global warming.

What's Next?

Despite its limitations and the challenges ahead, the Kyoto Protocol has already shown much promise, cutting 23% of CO_2 emissions from participating nations with emissions limits. The World Bank provides details showing that by 2015, we have seen a threefold increase over the past decade in carbon pricing instruments covering emissions, such as green bonds and carbon credits, translating to about 12% of annual global emissions.[49] Kyoto had a transformational effect in technology and emissions reductions and, at the same time, it was able to make a significant wealth transfer of above $215 billion through mid-2012 towards poor nations for clean and productive projects that reduced global emissions.[50] This transfer is desirable and fair. It bears repeating that the developing nations have emitted small amounts of carbon historically, and currently use little energy, while they bear a disproportionate burden of the risks of climate change. All of this was possible because of the magic of the carbon market.

Yet, there is more magic to behold. As we will soon see, the carbon market can avert global warming at no net cost to the global economy. It can foster sustainable development and close the global divide between rich and poor nations without cost to the taxpayer, while providing incentives to create and implement the clean technologies of the future.

The suspense in this drama was whether the carbon market could be fine-tuned and perfected in 2015. The COP21 could have turned the corner in 2015, but it failed to do so. What did COP21 in Paris achieve? The following chapter explains what was achieved and provides a map to the future.

4

The Road to Paris: An Insider's Timeline

The Kyoto Protocol was designed to be a first step, an experiment in how to reduce greenhouse gas emissions around the world by international agreement; it indicated the way forward. The 2015 global climate negotiations in Paris, known as COP21, created the Paris Agreement and demonstrated universal concern for climate change. Yet in 2017, the U.S. withdrew from the Paris Agreement.

The 1997 Kyoto Protocol took a long time to emerge and was the culmination of a long and contested process of information gathering and diplomatic negotiations for the nations of the world. The drama continues today. Not even the Treaty of Versailles, which ended World War I, or the Bretton Woods system, which rebuilt the world's war-torn financial system in 1944, were as long, complex, and difficult to negotiate as the Kyoto Protocol.

Good things take a long time to make. The history of the Kyoto Protocol and the Paris Agreeement is fascinating. It contains many elements of a blockbuster Hollywood film — suspense, drama, intrigue — and it is easy to lose sight of the seriousness of what is at stake. If climate disaster strikes in this story, it will not end — the cinema lights won't pop on and the credits won't start to roll.

Indeed, the Kyoto Protocol is the only international agreement we have with mandatory limits to avert catastrophic risks of climate change. It is a work in progress and going through growing pains. We can expect that international agreements like the Kyoto Protocol will not fully materialize for at least a decade. Dr. Chichilnisky's part in this process involved a long effort of many years in the diplomatic, scientific, and policy work that resulted in the creation of the carbon market of the Kyoto Protocol. This is the first time that an international agreement was created on the basis of a global market, the carbon market. This chapter will explain in some level of detail how this new and almost implausible achievement was realized. This is a "do it yourself" manual for creating international law. This chapter includes both aspects of international diplomacy and personal experience.

So far, the U.S. has refused to ratify and comply with the Kyoto Protocol, and in 2017 the U.S. announced that it was withdrawing from the Paris Agreement as well. Four years are needed to complete the withdrawal, so the matter will be decided by the next U.S. election. Under the 1992 United Nations Climate Convention, neither China nor India is obliged to curtail its emissions. They are poor nations, while most emissions historically originated from rich industrial nations. The U.S. views this as a major stumbling block to achieving sustainable emissions reductions, something that could undermine its own emissions reduction efforts. On top of this, there are fears in the U.S. of unfair advantage and competition for global leadership. China is the U.S.'s major global economic rival, following a decade in which China's economy expanded at a roughly 10% growth rate each year.[1] In the time since the Kyoto Protocol was initially designed and voted on in 1997, China has become a major economic power and is now the world's largest emitter of CO_2.

Developing nations view any demands on them to curtail emissions as unfair, since they are responsible for approximately equal shares of cumulative greenhouse gas emissions from 1850–2010 while they house about 83% of the world's population. Historically, most of the carbon emissions were produced by industrial nations in the process of industrialization.[2] At present, developing nations use energy more efficiently in terms of gross domestic product (GDP). GDP is tied to energy use, and reducing emissions means reducing economic growth, as Figure 1 "GDP and Carbon Emissions" illustrates.

Since the climate crisis cannot be solved without reducing carbon emissions, it is clear that the current impasse in global negotiations must be solved. The solution must include a clear timetable for a commitment from the industrial and developing nations to reduce emissions under some set of acceptable circumstances, in the future. But we may be getting ahead of the story. To understand the current situation, we must start from the beginning. How did it all start?

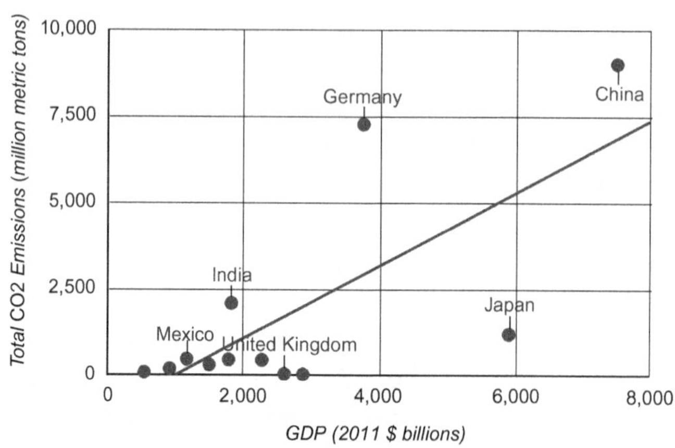

Figure 1 GDP and Carbon Emission
Source: The World Bank, http://data.worldbank.org.

An Insider's Timeline of the Kyoto Protocol

Dr. Chichilnisky was the architect of the Kyoto Protocol's global carbon market. She offers the following history and personal account of the Protocol.

Climatology was once a small and obscure branch of science. Yet important discoveries in the 19th century made it one of the most important fields of scientific study in the world. Listed below are some key dates in climate change history, and my personal account of the creation of the carbon market of the Kyoto Protocol. In 2015, the Paris Agreement only reached voluntary targets. At present, negotiators from over 195 countries are again attempting to draft a plan to solve global warming.

The scientific basis is laid

In 1824, the French physicist Joseph Fourier is the first to describe a "greenhouse effect" in a paper delivered to Paris's Académie Royale des Sciences.

In 1861, Irish physicist John Tyndall carries out research on radiant heat and the absorption of radiation by gases and vapors including carbon dioxide and water. He shows that carbon dioxide changes the atmospheric quality so that the atmosphere admits the entrance of solar heat but blocks its exit. The result is a tendency to accumulate heat at the surface of the planet.[3]

In 1896, Swedish chemist Svante Arrhenius first proposes the idea of a human-made greenhouse effect. He hypothesizes that the increase in the burning of coal since the beginning of industrialization could lead to an increase in atmospheric carbon dioxide and heat up the earth. Arrhenius was trying to find out why the earth experienced ice ages. He thought the prospect of future generations living "under a milder sky" would be a desirable state of affairs.

In 1938, British engineer Guy Stewart Calendar compiles temperature statistics in a variety of regions and finds that, over the previous century, the mean temperature had risen markedly. He also discovers that carbon dioxide levels had risen 10% during the same period. He concludes that carbon dioxide is the most likely reason for the rise in temperature.

In 1955, John Hopkins University researcher Gilbert Plass proves that increased levels of carbon dioxide could raise atmospheric temperature. (By 1959, Plass is boldly predicting that the earth's temperature will rise more than 16°C [28.8°F].) In the same year, chemist Hans Suess detects the fossil carbon produced by burning fuels. Although he and Roger Revelle, director of the Scripps Institute of Oceanography, declare that the oceans must be absorbing the majority of atmospheric carbon dioxide, they decide to conduct further research.

In 1957, a seminal article by Revelle and Suess reports "[H]umans are now carrying on a large-scale geophysical experiment," and in 1958, Revelle and Suess employ geochemist Charles Keeling to continuously monitor carbon dioxide levels in the atmosphere. After only two years of measurements in Antarctica, an increase is visible. The graph becomes widely known as the Keeling Curve, an icon of global warming debate and continues to chart the year-on-year rise in carbon dioxide concentrations to this day.

By 1963, the Conservation Foundation reports, "It is estimated that a doubling of the carbon dioxide content of the atmosphere would produce a temperature rise of 3.8°C (6.84°F)."

In 1979, NASA reports, "There is no reason to doubt that climate change will result from human carbon dioxide emissions, and no reason to believe that these changes will be negligible." Notice that this was almost 37 years ago and yet the problem is still with us today.

Climate change as a global concern emerges

In 1979, the first International Conference on Climate Change takes place; it involves mostly scientists. The First World Climate Conference introduces the threat of climate change to the global community and calls on the nations of the world to anticipate and guard against potential climate hazards. The World Climate Programme is established at this meeting. This is the first of many international conferences on climate change.

Political interest in climate change peaks after the 1985 Villach meeting in Austria, at which scientists at the World Climate Programme conference confidently predict that increased carbon dioxide concentrations will lead to a significant rise in the mean surface temperatures of the earth.

That same year, a hole in the ozone layer is discovered over Antarctica. It provides further evidence that human economic activity is altering the planet in dangerous ways. The 1980s would officially become the hottest decade on record, containing the four warmest years on record at the time, and since 2014 every new year has become the hottest year on record, further demonstrating these warming effects. The 1987 Brundtland Commission's well known report "Our Common Future," introduces the concept of sustainable development to the international community based on Chichilnisky's 1974 concept of *basic needs* and adds further fuel to the emerging climate debate.

The political agenda develops rapidly after two 1987 workshops sponsored by the Beijer Institute in Vilach and Bellagio. It becomes clear that in order to address the climate problem the scientific issues have to be clarified first. This is achieved by an interdisciplinary group of scientists across the world that includes physicists, atmospheric scientists, biologists, and economists, all working as part of the Intergovernmental Panel on Climate Change (IPCC), which is set up in 1988 by the World Meteorological Organization (WMO) and by the United Nations Environment Programme (UNEP). The IPCC provides reports based on scientific evidence that are widely regarded as reflecting the dominant viewpoints of the global scientific community.

In 1988, a drought in the U.S. reduces parts of the Mississippi River to a trickle and sets much of Yellowstone National Park ablaze. The 1988 drought leads to a reduction in crop production of almost $20 billion and increases in food prices of more than $12 billion, along with $1 billion in increased transportation costs.[4] This further kindles public interest in to the possibility of climate change. In June, Dr. James Hansen of the NASA Goddard Institute for Space Studies at Columbia University delivers his testimony to the U.S. Senate. Based on computer climate simulations and temperature measurements, he states that "the probability of a chance warming of that magnitude is about 1 percent," indicating that the greenhouse effect has been detected and it is already changing the climate.[5]

In 1989, the Second World Climate Conference is held in Geneva, only this time it includes representatives from countries worldwide as well as members of the scientific community. It is this conference that lays the groundwork for the current international climate regime by calling for the creation of an international convention on climate change. The United Nations soon establishes a committee to negotiate the convention's text. It eventually becomes the United Nations Framework Convention on Climate Change (UNFCCC), the international body charged with the clarification, negotiation, and resolution of climate change. Not only is the UNFCCC notable as one of the first major international environmental agreements, but it firmly embraces the notions of sustainable development and shared responsibility for the health of the planet. In 1990, the IPCC delivers its first assessment on the state of climate change, predicting an increase of 0.3°C (0.54°F) each decade in the 21st century, greater than any rise seen over the previous 10,000 years. This report is influential in shifting public opinion in favor of the seriousness of climate change.

International Negotiation Begins in Earnest

In 1992, the United Nations Conference on Environment and Development, better known as the Earth Summit, takes place in Rio de Janeiro. It is attended by 172 countries. A total of 150 nations agree that the international community should focus on efforts towards "sustainable development," which are defined as those policies that satisfy the basic needs of the present without undermining the rights of future generations to satisfy their needs; I had introduced the concept of basic needs as part of the Bariloche Latin American World Economic Model in 1977.[6] The Earth Summit marks the first unified global effort to come to grips with global warming. Most notably, the UNFCCC is signed by 154 nations at the Earth Summit. Negotiations that take place at the Earth Summit eventually lead to the 1997 Kyoto Protocol.

Today, the UNFCCC enjoys almost universal membership. More than 196 countries have ratified it. Within the Climate Convention, the annual Conferences of the Parties to the Convention (COP) play the most important role. The COP is responsible for meeting and making the necessary decisions to implement the UNFCCC's objective. That objective is to stabilize greenhouse gas concentrations so as to prevent dangerous interference with the climate system and enable economic development to proceed in a sustainable manner.[7]

The UNFCCC affirms the global community's intent to preserve the climate system for present and future generations and set forth the principles of cooperation between states. It establishes the precautionary principle as the defining reason for global action to prevent climate change. It recommends the creation of cost-effective mechanisms for achieving emissions reduction, consistent with sustainable development and designed to provide "no regrets" safeguards against such risks. These steps should also be compatible with food security, social justice, and the wealth of nations.[8]

Most importantly, Article 4 of the UNFCCC establishes the notion of common but differentiated responsibilities for mitigating climate change on the basis of countries' contributions to the buildup of greenhouse gas concentrations in the past and their ability to afford reductions in the present. As we will see, it is this principle, founded in the spirit of fairness and partnership, that leads to serious conflict between industrialized and developed countries over whether and how to cap developing country emissions. It is this conflict, more so than any other, that underlies the future of the Kyoto Protocol.

Article 4 of the 1992 UNFCCC assures that industrial nations take the lead in emissions reduction and that developing nations will not be asked to reduce their emissions without compensation. All countries that participate in the U.N. Climate Convention have certain obligations. They must all provide greenhouse gas inventories, national strategies, measures and reports. The UNFCCC encourages industrial nations to reduce their emissions and to provide financial assistance to developing nations to achieve the conventions' goals. Very soon after the U.N. framework convention is completed, however, it becomes clear that most countries are not on track to meet their non-binding emissions aims. It is evident that a new agreement has to be negotiated.

It is worth stepping back for a moment to summarize how global climate negotiations operate, if for no other reason than to illustrate how international change processes unfold, how difficult it is to gain consensus and how necessary global cooperation is to achieving a solution. Through the use of framework conventions and protocols, the approach is to allow states to proceed incrementally; a framework convention establishes a system of governance, and specific obligations are developed in protocols. After the 1992 Earth Summit, the UNFCCC is created to establish a general system of international governance for climate-related issues. To build scientific consensus step by step as well, the work of the UNFCCC is based on the IPCC, the scientific body that includes thousands of scientists from all participating nations. The IPCC confirms the human origins of climate change in its second assessment report in 1995. Through the work of the UNFCCC, on December 11, 1997, 160 countries vote in favor of the Kyoto Protocol. But we are getting ahead of the story…

Between the Earth Summit in 1992 and the creation of the Kyoto Protocol in 1997, several notable events occur. First, the UNFCCC enters into force in March of 1994 and the COP becomes its ultimate authority. The COP convenes for its first meeting in Berlin in 1995. The IPCC, the scientific advisory board to the UNFCCC, completes its second assessment report in 1995 just in time for the second COP meeting in Geneva. This report issues the first official statement confirming humans' impact on the global climate, finding a "discernable effect of human carbon emissions on the earth's climate." As if to drive home the message, temperatures around the world soar to record highs in 1995. The 1990s soon replace the 1980s as the hottest decade on record. Global temperatures continue to trend upwards in a significant manner, leading to new record highs consistently as time progresses, with 2005–2015 becoming the hottest decade on record.[9] There is no longer much doubt that climate change is a real risk with a human cause. Public opinion worldwide wakes up to the realities of climate change and demands solutions.

Emphasis shifts from science to economics

By the early 1990s, the international community realizes that the key to resolving the physical problem of climate change lies in social organization, in particular in economics. The causes of climate change are economic, steeped in the ways we use energy to produce goods and services in the world economy, while the effects are best measured and understood by the physical sciences. As scientific consensus begins to emerge about the human causes of climate change and its possible effects, the emphasis shifts from the science of climate change to its economics. This complicates the problem enormously, as these are two disciplines that have rarely communicated before. The effects of climate change are physical; therefore economists are at odds to measure them. Yet the causes of climate change are economic; therefore there is little or nothing that physicists can do to resolve the problem on their own. The IPCC realizes that it needs to enlarge its work by including economists.

I have been concerned about the global environment since I created the Bariloche model in the 1970s, during which I introduced the concept of *basic needs* that was voted on by 150 nations at the Earth Summit in 1992 as the main concept for economic development. *Basic needs* is another form of defining economic progress, different from GDP, which I created in the mid-1970s as a foundation of the Bariloche model and in response to the Club of Rome's *Limits to Growth* model of the world economy developed at MIT by Daniel and Donella Meadows. It was the first formal attempt ever to model how resource limits would limit global economic growth. The Bariloche model offered a response to the *Limits to Growth* assertion that the developing nations could not grow without destroying the world's resources. Bariloche made the argument that developing nations could grow if they focused on the satisfaction of *basic needs*, which is the economic concept that I defined. In 1992, *basic needs* became the corner stone of efforts to define sustainable development by the Brundtland Commission at the 1992 U.N. Earth Summit in Rio de Janeiro. Sustainable development is not defined by GDP. Instead, it is an economic strategy that satisfies the *basic needs* of the present without interfering with future satisfaction of needs. After the Rio event, *basic needs* became widely known and adopted by U.N. organizations around the world, including many country case studies by the International Labor Office in Geneva. I decided that the U.N. had the capacity to implement this new way of defining economic progress that was harmonious with the global environment, and started working as the director of the U.N. Institute for Training and Research project on technology and *basic needs* under the aegis of Phillippe de Seynes, former Under Secretary General of the U.N. for Economic and Social Affairs. This started my work of several decades with the United Nations. By the mid-1990s, I started collaborating with the IPCC, and by 1996 I became a lead author of the IPCC, representing the U.S. It is in this role that I started my work in the UNFCC that led to the creation of the Kyoto Protocol and its carbon market.

The increasing emphasis on the economic issues of developing nations in that decade also leads to the recognition that a crucial aspect of the environmental problem lies in the relationship between rich and poor nations. While most of the emissions originated historically from the rich nations, in the future the developing nations could become the overwhelmingly largest emitters in the world. The concentration of carbon in the atmosphere of the planet is the same for all nations. Each nation can inflict damage on the rest by emitting carbon. Therefore, all nations must agree to reduce emissions. Collaboration between the rich and the poor nations becomes central to resolving the climate problem. But it is a thorny issue, since developing nations need energy to develop, and industrial nations worry that growing energy demands in the developing world will further fuel global warming. Without collaboration between the two groups of countries there can be no solution.

In June 1993, the Organization of Economic Co-operation and Development (OECD), an international organization that represents the rich nations of the world, hosts an international conference at their offices in Paris, France, in which the major players in the global climate negotiations explore the connection between global warming and the economy. (In retrospect, this conference laid the groundwork for the framework of the Kyoto Protocol in 1997.) The conference includes the Economics Division of the OECD, the department where the economists Peter Sturm and Joaquim Oliveira Martins work. It includes representatives from industrial and developing nations, in recognition of the need to involve all nations in the resolution. One of the speakers is Ambassador Raul Estrada Oyuela, who later becomes the lead negotiator of the Kyoto Protocol. Another participant is Jean-Charles Hourcade, who becomes part of the French delegation to the UNFCCC, and who later invites me to write the crucial wording introducing the carbon market into the Protocol.

In my official presentation at this conference, I propose to the OECD the creation of the carbon market that eventually becomes an essential part of the Kyoto Protocol. I explain the value of the market approach and the need for preferential treatment for developing nations, because the concentration of CO_2 in the atmosphere is a global public good, effectively establishing the connection between equity and efficiency in carbon markets: the carbon market has the potential to become an effective way to promote efficient allocations of resources in the economy (efficiency), encouraging clean technologies, while at the same time reducing the inequalities of income and consumption throughout the world economy (equity). The proposal I present at the 1993 OECD meeting eventually becomes the foundation for the creation of the carbon market and the Clean Development Mechanism (CDM) of the Kyoto Protocol. But at this point in time, my presentation is controversial and leads to much debate. Raul Estrada Oyuela, who represents Argentina in the global negotiations, is against the use of markets in environmental conservation, as is Peter Sturm, who opposes the results linking equity and efficiency in carbon markets. Most European economists at the time are against the market approach and prefer carbon taxes, a political issue that is discussed in more detail in this book. The debate on carbon markets arising from this conference grows surprisingly large, and involves economists and diplomats across Europe, the U.S., and Latin America. I publish several articles and books showing, for the first time, that in economic terms carbon concentration in the atmosphere is really a global public good and, therefore, its properties are quite different from standard private goods.[10] In particular, I show that markets with global public goods such as the planet's atmosphere are quite different from conventional markets, and it is impossible to separate efficiency concerns from equity concerns, as is the case with private goods.[11]

The controversy grows...

At this time, almost everyone believes that the majority of the abatement of carbon emissions should be done in developing nations, based on the assumption that the costs would be lower. Economist Larry Summers famously argues in an internal World Bank memo that developing nations should be sent the pollution from rich nations, as it would be more cost efficient for them to clean pollution. Geoffrey Heal and I then publish an article in Economic Letters in 1994 showing exactly the opposite, namely that abatement should be done mostly by industrialized nations, that it is more efficient if rich nations abate their own emissions even if it costs more. This fuels the controversy further since the conventional wisdom that applies to private goods, but not to public goods, dictates the opposite.

As the controversy grows on both sides of the Atlantic, my research and my proposal for a global carbon market become well-known sources of scientific debate. The OECD starts to recognize the diplomatic and economic importance of the issues involved, and I become a consultant, writing a report on carbon markets for the OECD Economics Division with Professor Heal.[12] I also continue working as an adviser to various U.N. organizations and the World Bank, as I had done since the mid-1970s. In that role I build with Heal at Columbia University a version of the OECD Green Model of the world economy and enlarge it to include a carbon market. This becomes the first model of the global economy that simulates the behavior of the global carbon market. I publish the results in a book produced under the auspices of the U.N. entitled, *Development and Global Finance: The Case for an International Bank for Environmental Settlements.*[13] These results show the efficiency advantages of assigning the poor nations more rights than rich nations to produce greenhouse gases.

In 1995, I publish "Markets for tradable CO_2 emission quotas: principles and practice" with Heal, under the *OECD Economics Department Working Papers*, No. 153. For the first time, it introduces the idea of the global carbon market to a major international organization led by industrial nations: the OECD. We explain how the global carbon market could operate in practice. This is the beginning of the carbon market's public role in Europe. We also explain the advantages of markets over taxes for reducing global emissions.

It is critical to understand the connection between equity and efficiency because it is linked to political issues between the OECD and the developing nations. The controversy arising from my OECD presentation and the OECD report that we produced, about how equity was needed for the efficiency of the carbon market, continues to grow, involving Peter Sturm, head economist of the Economics Division of the OECD, and other distinguished economists in the U.S., such as David Starrett, who was then the Chairman of the Economics Department at Stanford University. Heal and I write a joint paper with Starrett on the basic economic structure of the global carbon market, laying down the rationale for a preferential role for the developing nations. Sturm and his colleague Joaquim Oliveira Martins write a piece explaining their different position on the matter, in which they claim that equity and efficiency are disconnected. Eventually it becomes clear that the difference was due to their assumption that environmental quality is not a factor in economic welfare. Heal and I decide to memorialize the entire debate, including all authors, warts and all, in a book that is published in 2000 by the Columbia University Press, *Environmental Markets: Equity and Efficiency*.

During 1994 and 1995 I continue presenting my case for a carbon market proposal in many universities in the Americas, Europe, Asia, and Australia, as well as at the U.S. Senate and U.S. Congress. My case has two parts. The first part argues for the market because it starts by requiring global emissions limits nation by nation. The second part, which has to do with "equity," recognizes that since carbon concentration in the atmosphere is a global public good, it requires a Lindahl solution for efficiency and thus, in turn, implies a measure of equity in the treatment of the poor nations. The first part is congenial to mainstream economists in industrial nations, while the second is attractive to environmentalists and politicians from developing nations. While market efficiency is never debated, the role of equity in attaining efficiency remains a continued uphill debate. There is some degree of sympathy in giving a preferential role to poor nations, but it stems from a sense of charity or goodwill rather than a desire to secure economic efficiency. The reality is that the global carbon market trades a very different type of good, a global public good, and is therefore a different market from any other in the world. The carbon market represents a new type of economics.

As time passes, I press my market approach in academic and political publications in newspapers and magazine articles and even on TV. In 1994, I publish an article in the *American Economic Review* (*AER*), "North–South Trade and the Global Environment," showing how the global warming problem and the global poverty problem are one and the same. Both derive from excessive exports of natural resources by developing nations. Developing nations do not export natural resources because of comparative advantages. More often than not, they over-extract and export natural resources because they lack private property rights over those resources. Resources in developing nations are often treated as common property and used on a first come, first served basis. As I show in the 1994 *AER* publication, this leads directly to a global tragedy of the commons: overexploitation of resources in the South (developing countries) and overconsumption in the North (developed countries). I explain in this article how the excessive emphasis on resource exports, such as petroleum, damages the economy of developing nations and undermines their ability to develop, grow, and feed their people. The Kyoto Protocol carbon market is designed to correct all of this: it creates property rights on the use of global commons, resolving the tragedy of the commons. These results are considered almost heretical at a time when most economists are espousing the then popular theory of "export-led growth." Now, the results are widely accepted and often considered self-evident.

At the same time, I argue for the equity features that need to be included in global carbon markets as an advisor to the U.S. government, and various U.N. organizations, and as a lead author of the IPCC. My article with Professor Heal in Economic Letters presents the scientific case for a link between equity and efficiency in these markets, as do our OECD 1995 report and the Pegram Lectures I give later at the Brookhaven National Laboratories on Long Island, New York, in 1999.[14] The equity-efficiency link in global carbon markets becomes my trademark — and a continued source of debate on both sides of the Atlantic and in international organizations.

To drive the point home and show how the carbon market would operate, I develop, with Heal and Yun Lin, a computerized model of a world economy that includes a global carbon market at the Program on Information and Resources (PIR/Green) at Columbia University. This new global model extends the earlier OECD GREEN global model. It introduces a market for emissions trading and shows in practical terms the positive impacts that a global carbon market can have on the global environment and on the world economy.

The empirical model is useful in helping people visualize the creation of the new global carbon market and how it would operate. This is the first time a global economy is seen functioning with a global carbon market, and the benefits become apparent. I present the results in the U.S. Congress and at several U.N. meetings. The results show that policies that benefit developing nations — for example, mixed allocations of permits that include population as well as GDP as bases for the allocation — are more efficient than those that simply consider GDP. "Grandfathering" (awarding emissions rights to nations according to their own historic pattern of emissions) is, in fact, less efficient than providing more allocations to the poor nations.[15]

The global model emphasizes the value of allocating more emissions rights to developing nations and explains how this would work in practice. Our model creates a blueprint for the allocation of emissions rights that eventually prevails in the Kyoto Protocol in 1997.

Kyoto begins to take form

Until the mid-1990s, the public is still unconvinced of the risks of climate change; the issue is still widely disbelieved and misunderstood. However, by 1995, the hottest year on record, the discourse begins to shift. The second IPCC Assessment that year states for the first time that "the balance of evidence suggests a discernible human influence on global climate." This is the first scientific assessment that gives an unequivocal signal that the earth's climate is changing and its changes are primarily due to human causes. The international political response is to take the issue of climate change more seriously and to seek global solutions.

In December 1996, I am invited to give the keynote speech at the Annual Meeting of the World Bank in Washington, D.C. I officially propose the creation of a global carbon market, explaining why this solution would be superior to carbon taxes and how this approach could favor both industrial and the developing nations. (The difference between the carbon market and the carbon tax was the subject of an OECD report I wrote with Heal in 1995). This presentation is attended by hundreds of people, including my colleague Sir Partha Dasgupta, who comments very favorably. It becomes the first official U.S. proposal for the carbon market and its CDM.

In 1996, the UNFCCC Convention of the Parties meets in Berlin and agrees on the so-called Berlin Mandate, which is, one might say, an "agreement to agree." This Mandate requires U.N. negotiators to come up with a solution to the climate change issue by the next COP meeting in Kyoto 1997. Ambassador Raul Estrada Oyuela is elected as the lead negotiator for the negotiations in Kyoto, and he inherits the same mandate: to reach an agreement at the next COP. As a professional diplomat, Raul takes the mandate very seriously, and is determined to reach an agreement in Kyoto. The task is enormous, since industrial and developing nations are now more divided than ever on the climate change issue.

Believing that the only way to reach an agreement is to find a solution that appeals to both industrial and developing nations, I continue to press my case in favor of the carbon market in presentations and publications around the world. My proposal for a global market approach protects developing nations and elicits a favorable response from the U.S., which supports market approaches. Any proposal that has the simultaneous approval of the industrial and the developing nations is a good start, since the most thorny and challenging aspect of the climate negotiations is the tug of war between the rich and the poor nations.

I label my global carbon market proposal "a two-sided coin" because it offers a market solution that addresses the industrial nations' concerns for efficiency and flexibility, while simultaneously offering developing nations an "equity" approach through the allocation of rights to emit that appeals to their own concerns for poverty alleviation and historical fairness. My own view in debating with the U.S. and developing nations' representatives is that the positions of the global North and the South are so diametrically opposed that only a solution that appears opposite to each side will prevail. My proposal for a global carbon market truly looks opposite to each side. However, in all honesty, my carbon market approach is not popular at the time. In fact, for many years, it is opposed by all sides. Neither developing nations, nor environmentalists, nor industrial nations favor it. Academics are also surprised or antagonized to hear my arguments that a global carbon market would blend equity with efficiency; they are skeptical that a market could genuinely achieve both. At the time I am a Trustee of the Natural Resources Defense Council where Robert Kennedy Jr., the son of President Kennedy's brother, is an environmental attorney. On two occasions I debate the issue with Kennedy Jr. at the Columbia University Reuters Forum on TV; his telegenic anti-market environmentalism carries an easier emotional appeal than my rational pro-carbon market approach, although there is no essential difference in the end goal, once it is understood that markets must change in order to be harmonious with the environment. His argument is simple: the market is the enemy of the environment so how can we enlist the market to solve the largest environmental problem of our times? It is true that the market has caused many of the problems we face with the environment. At the same time, the change in markets proposed in this book is what reconciles these two approaches. We need to change the markets, not ignore them.

Kennedy Jr. is not alone. Raul Estrada Oyuela, a great environmentalist who had become a collaborator of mine in conferences and writings since the 1993 OECD meeting, as well as the Lead Negotiator of the Kyoto Protocol, is also set against the carbon market. This is more serious. Indeed, none of the environmentalists I know at this point are in favor of the carbon market. They express that the idea is like "trying to buy or sell your own grandmother — physically possible but morally repugnant."

It will take all of my 20 years of credibility as a developing nation supporter and as the creator of the concept of *basic needs* to get developing nations to support the idea. My explanation that the carbon market is different from all other markets because of the global public good aspects of atmospheric carbon dioxide, which induces an equity principle in the way emissions limits are assigned, is somewhat helpful. But it is a hard sell, as it is a difficult concept to comprehend. In reality, because the carbon market trades a global public good, it is very different from all other markets we know. It is unique in history.

The Europeans in the OECD are also dead set against the carbon market in 1997. Their approach is closely aligned with carbon taxes. The late James Tobin, a great U.S. economist and a Nobel Laureate from Yale University, publicizes his proposal for a general approach to global warming, based on creating a global carbon tax for reducing carbon emissions. His proposal is called "The Tobin Tax." Neither the global North nor the South, namely the rich and the poor nations, nor most of my own fellow academics, are on my side in this debate. Once again, just as it was when I defined *basic needs*, I find myself arguing against the generally accepted positions, the mainstream point of view. This time it is even more challenging, because my views also conflict with those of mainstream environmentalists.

Arguing my case

Paradoxically, I find myself arguing for a carbon market solution to which environmentalists are opposed. I am advocating for a market solution, the global carbon market, which opposes the tax approach popular at the time within the industrial nations, the U.S., and the OECD as a whole. In doing so, I oppose Yale University economist, James Tobin, who was well known and liked. Tobin advocates for a global tax on carbon while I advocated for fairness toward developing nations, proposing a free market approach. The paradox is more illusion than fact. The carbon market starts with setting firm ceilings on emissions and is therefore closer to the heart of an environmentalist than a carbon tax, which has no certain impact on the quantities emitted. It allows each nation to use their own internal approach, such as limits, markets, or taxes. Eventually, despite the almost universal feeling that markets favor industrial nations and are dangerous for the environment, I am somehow able to turn the tables and show that the environment and the developing nations will benefit more from the carbon market than from using carbon taxes (for more on why this is, see Chapter 5).

In 1997, in addition to participating in seemingly unending academic and U.N. debates, I present my proposals through the U.S. Under Secretary of State, Timothy Wirth, to experts in the U.S. Department of State, and through the U.S. Under Secretary of the Treasury, Larry Summers, to experts in the U.S. Department of the Treasury. Both Wirth and Summers are interested and supportive. Naturally, they are my supporters, possibly the only supporters and allies I have at this point. However, I suspect that they could be supporting my ideas for the wrong reasons. They do not seem to understand, nor do they appear to care much about, the 'equity' piece of the carbon market that is so important to me, and, of course, to developing nations. Indeed, the debate on this North–South equity issue in the carbon market still continues today. Yet, eventually the two-sided coin argument prevails. The attractiveness of the market approach wins the day in the U.S. I attract attention to the topic in every possible way I can, including in the U.S. Congress and Senate, where I receive interested, if slightly unfocused, attention. In the process I meet with U.S. Vice President Al Gore, a very intelligent man with a positive agenda but who is not inherently interested in market approaches.

I organize special meetings at Columbia University with the negotiators of the Kyoto Protocol, to explain the somewhat counterintuitive fact that the market approach will benefit the developing nations of the world. In the autumn of 1996, Peter Eisenberger creates the Earth Institute at Columbia University, based on the research agenda of the Program on Information and Resources (PIR) that I founded and have directed since 1994. Eisenberger views my efforts as the cornerstone of the research agenda on globalization at Columbia University. PIR starts a close cooperation with the Earth Institute, which joins forces with my proposal for a global carbon market. This strategic alliance between the Earth Institute and my center, PIR, at Columbia University proves very productive, and becomes a crucial step in the road to Kyoto. The lead negotiator for the Kyoto Protocol, Raul Estrada Oyuela, becomes a distinguished visitor to PIR for a year. At PIR, I develop the OECD Green Model of the world economy to include the carbon market, in order to test its impact.

In April 1997 I publish my book *Development and Global Finance*, invited by the United Nations Educational, Scientific and Cultural Organization (UNESCO) and United Nations Development Programme (UNDP), exploring how the carbon market works and how the same market approach can solve other environmental issues, such as biodiversity destruction, and water scarcity. The book reports on the results of the PIR/Green (a modified OECD) model, showing how equity benefits efficiency in the context of the carbon market. In January 1997, I publish a *Financial Times* article, "The Greening of the Bretton Woods" where I attribute the origins of the global warming problem to the global trade of resources following the creation of the Bretton Woods institutions. Here, once again, I propose the creation of the carbon market.

My *Financial Times* article has a larger impact than I could have fathomed. It attracts the interest and support of Mohamed El Ashry, then the Director of the Global Environmental Facility at the World Bank. The reaction to this article leads me to organize, along with Columbia's Earth Institute, a small but important meeting at the Rockefeller Foundation's Bellagio Center on the shore of a beautiful Italian lake. The meeting focuses on my proposal for an International Bank for Environmental Settlements, and includes the movers and shakers of the time, such as Tom Lovejoy who later led the Heinz Foundation and Hazel Henderson, a brilliant economist, futurist, and creator of the Ethical Markets Media television series. The purpose of the meeting is to execute the plan for creating a carbon market, along wtih an International Bank of Environmental Settlements to support its banking aspects, such as the borrowing and lending of carbon rights. In mid-1997, I become a lead author of the IPCC Working Group II, and attend several meetings for their next Assessment Report. The work we create in 1997 becomes the basis for the Nobel Prize that was awarded 10 years later to the IPCC and U.S. Vice President Al Gore in 2007.

In 1997, I also work with Exxon representatives, trying fruitlessly to change their views about whether the global warming threat is real. The same year, the Earth Institute hires a publicist named David Fenton, who later goes on to aid the rise of moveon.org, a famous political website. At the end of the year, the Earth Institute sends a delegation to Kyoto for the December 1997 COP meetings. I am part of the Earth Institute delegation together with its director Peter Eisenberger and faculty member Rick Fairbanks, both famous physicists. Eisenberger is at this time Vice Provost at Columbia University, Founding Director of the Earth Institute and Director of the Lamont Doherty Earth Observatory at Columbia University. He supports me but is against the carbon market, which he views as a child of capitalism, more a part of the problem than of the solution.

When we arrive the city of Kyoto is grey and the weather is humid and sticky. The large tents where the Convention of the Parties meets are buzzing with people and heated with debate. There are hundreds of earnest non-governmental organizations who participate in the COP meeting, spilling over in to the streets near the U.N. convention and throughout this beautiful Japanese city. I continue with an unending stream of presentations, press conferences, and interviews with the global media on my carbon market concept, which is attracting a lot of attention. These interviews are published in a number of newspapers worldwide and appear on several radio and TV shows. I give a "side event" presentation at the COP that is very well attended. Here is a brief description: it all happens under a huge white tent, like a circus, and the floor is covered with wood shavings. There are journalists everywhere and people from all over the world carry signs expressing concerns for the environmental disaster they see coming. Brian Flannery of Exxon is also there, and continues his attacks on the whole exercise. (More recently, he exposed Exxon's own position in denying the problem of climate change, which prompted several U.S. states to sue Exxon.)

It seems fair to say that people do not really understand much of what I am advocating for and agitating about, although they are inclined to give me a hearing. I am basically working alone in this task, trying to create the carbon market of the Kyoto Protocol. Except for my colleagues at the Earth Institute, who are in Kyoto with me, and several of my UNFCCC colleagues, such as Raul Estrada Oyuela and Jean-Charles Hourcade, both of whom had worked with me since the OECD 1993 meetings in Paris and now had official roles as the Lead Negotiator and French representative respectively, I do not know many people. However, I am known to and on a first name basis with many of the official negotiators of the developing nations. This is how I lobby, I talk to all the U.N. representatives at every occasion I can, particularly Oyuela, who, as Lead Negotiator, is getting more and more worried about the ability to reach an agreement but is adamantly against the carbon market approach, and Hourcade, who is rather young and willing to hear my carbon market enthusiasm, apparently on the strength of my work in international trade and social choice, which he professes to admire. I have lunch with Amory Lovins and with other well-known U.S. figures from the energy industry. My colleague William Nordhaus from Yale University is also there, as is James Cameron, a brilliant British attorney who represents the group of island nations of the world in the UNFCCC Kyoto Process. Also present is Ambassador Dasgupta, who represents India in the negotiations, the IPCC Chairman Robert Watson, who later becomes Senior Scientific Advisor on the Environment to the World Bank, Urs Luterbacher, a reputed political scientist from Geneva, Jacques Weber, an excellent French economist, and Kilaparti Ramakrishna, an environmental attorney who advises Oyuela and goes on to become a Senior Advisor on the Environmental Law and Conventions at UNEP in Nairobi.

At the end there is a crisis: the feared deadlock between the industrial and the developing nations, the North and the

South. The same problem exists today and is the cause of the impasse between the U.S. and China. I am acting as an unofficial adviser to the UNFCCC COP in Kyoto and I am also well recognized in my official role as a lead author of the IPCC. It is in this role that I gather psychological strength and address Oyuela head on. He is friendly and willing to hear me out. I tell him why the carbon market approach is like a two-sided coin, and the only possible solution to the dilemma he faces, because it could be supported by both the industrial and that the developing nations. I explain that the U.S. will view it favorably because it is a market approach and that the developing countries will approve of it because it sets mandatory limits on the rich nations' emissions and not on poor nations. Instead of mandatory limits, the carbon market will give the poor nations strong incentives for reducing emissions, along with the CDM, by which funding from the carbon market can be used to support projects in developing nations that decrease their emissions below certain agreed milestones or baselines. There are precedents in the U.S. for the use of markets to resolve environmental issues, but none at the global level. The sulfur dioxide (SO_2) market in the Chicago Board of Trade of Illinois is an excellent prior example, but the SO_2 market is very different from the CO_2 market because CO_2 spreads uniformly around the world, making the carbon market trade a public good, which is not the case with SO_2. A carbon market creates a price for carbon such that emitting CO_2 would be costly, and reducing emissions would be profitable.

Moreover, I had proven in my publications that trading in a global public good would make the carbon market require some form of equity in the distribution of emissions rights to favor poor nations, justifying an approach favorable to developing nations and encouraging the creation of the financial mechanism of the CDM. These facts had been the essence of my 1993 presentation at the OECD and of several academic articles I had published by then, validating the equity principles behind markets trading public goods, and requiring a favorable treatment for poor nations. This had been explained in my 1993 presentation at the OECD, *The Trading of Carbon Emissions Rights by Poor and Rich Nations* and in my book with Heal, *Environmental Markets: Equity & Efficiency*.

On the lobbying front, Oyuela listens to me with interest and sympathy but remains emotionally set against a market approach. Oyuela is a lawyer, and for him mandatory emissions limits are the only possible solution. The idea of trading the rights to emit between nations — that is, the nations who exceed their limits can buy rights to emit from those who are below — seems to him almost unethical. He understands and sympathizes with the Clean Development Mechanism (CDM), a strong incentive system that favors developing nations, since poor nations use little energy and are below their projected emissions limits, meaning they will receive monetary compensation from the CDM via the carbon market's funds. Nevertheless, I insist that the carbon market and its CDM are the only possible solution for climate change, since they simultaneously provide mandatory emissions limits for the rich nations, who generate the bulk of global emissions, while creating strong incentives for poor nations to reduce their own emissions. Furthermore, I explain to Oyuela that I had advocated and agitated enough, for several years, with the developing nations (known as the Group of 77, or G77) negotiators to know that they would support the carbon market.

I had discussed with and given official presentations at the U.S. Senate and Congress. I had also discussed with Larry Summers, who was Bill Clinton's Under Secretary of the U.S. Treasury, and with Timothy Wirth, then Under Secretary of the U.S. Department of State, and therefore knew that the carbon market had sympathizers in the U.S. government as well. The carbon market would effectively allow the critical mandatory limits on emissions that scientists required, while offering a bit of flexibility for the rich nations to satisfy their quotas. Oyuela listens. I explain that the only remaining thorn is Europe. That is where I had done less work and Europeans are more adamantly against markets, which they view as a new-fangled U.S. excuse or trick to avoid abiding by carbon emissions limits. The Europeans favor taxes, but taxes do not offer mandatory limits and there is no world tax authority to impose them, so they presented more of a fantasy than a reality. Having explained all this to Oyuela, his response is: "we will see." To his credit, he considers any position that is proposed to him and evaluates it fairly. And he is a remarkable negotiator.

As the meetings progress, it becomes clear that the negotiators are deadlocked. The industrial nations do not want mandatory limits on their emissions, since most of the energy they use is fossil fuel based and they have no intention to sacrifice economic growth. Developing nations are even more adamantly opposed to accepting emissions limits, which they see as unfair for historical reasons, since the developed nations were responsible for most of the emissions up until then. They see emissions limits as a way to condemn their people to poverty and the trap of underdevelopment.

Result!

On December 10, the last day of the conference, the general mood is somber. Along with many others who understand the seriousness of the situation, I stand in the area just outside the large negotiation theater quite late at night, expecting to hear of a breakthrough that does not happen. It is truly sobering. Then at about 10 p.m., Jean-Charles Hourcade comes out of the negotiating room, which has graded seating like a theater, and invites me in.

Professor Hourcade is a well-known, highly regarded economist and French government official. He is an intelligent and original thinker and the author of one of my book's Preface. Since the 1993 OECD meeting, he had known about my proposal for the carbon market and the innovation that this would entail, in the sense of trading a global public good, namely the carbon concentration in the planet's atmosphere. He had known about the attendant issue of equity and efficiency involving the industrial and developing nations; he had accepted the consistency of this issue with Article 4 of the 1992 Climate Convention, which gives developing nations a preferential role.

Hourcade asks me to write a simple description of the carbon market to include in the draft of the Kyoto Protocol. He is one of three members of the contact group between the EU and the U.S. and needs me to prepare wording that will help the Europeans agree with the market approach, and provide the flexibility that the U.S. requires before accepting emissions limits. It is truly the eleventh hour of the Kyoto negotiations, because the meetings are ending on December 11, the very next day.

The agreement is difficult to achieve, particularly with the U.S., and the introduction of the carbon market is invaluable in reaching it. This is where my work of several years — developing the carbon market, at the U.N. and the OECD, making presentations to the U.S. Congress and Senate, the World Bank, and the General Agreement on Tariffs and Trade (GATT); and proposing a market structure to the U.S. Treasury and Department of State — actually pays off.

It is 11:00p.m. and, at Hourcade's request, I sit down on the steps of the negotiating room, and begin to write. I write, "Each nation has a mandatory emission limit. In a given year those nations that go above their limit can buy the rights to emit from nations who go below their limits. This way global emissions always remain below the required limits." The simplicity of my words makes the market acceptable to the European nations, becoming the basis of Article 17 in the Kyoto Protocol, which describes emissions trading and how the COP will develop the carbon marketing mechanism. In reality, the introduction of the carbon market saves the day. It creates a measure of flexibility that leads the U.S. to sign the Protocol. The U.S. feels it can trade its way through the problem: if it cannot meet the limits, it can buy rights to emit from other nations. As the developing nations have no limits on emissions, they have preferential treatment, so they sign as well. According to Hourcade, my role is critical to convincing the U.S. and European representatives who then sign on to the Kyoto Protocol. The introduction of the CDM provides the added flexibility to integrate developing nations into the carbon market framework, so they can benefit from technology transfers from industrial nations without facing emissions limits.[16] The 160 nations' representatives in Kyoto thus reach a historic agreement at the 11th-and-a-half hour. On December 11, 1997, the Kyoto Protocol is born, voted for by 160 nations, including the U.S.

Industrialized countries agreed in the Kyoto Protocol to cut their emissions of six key greenhouse gases by an average of 5.2%, a triumph for Ambassador Oyuela. Under the terms of the agreement, each country, except for developing countries, committed to reducing emissions by a certain percentage below their 1990 emissions levels by the period

2008–2012. The Kyoto Protocol is signed in 1997, and in a private discussion at Oxford University in the U.K. with former President Bill Clinton, I found out the historical circumstances under which this happened. I met President Clinton in 2013 at an exclusive conference organized at Oxford University by Baron de Rothschild, and he explained to me in detail what happened. He said that the Kyoto Protocol was signed by the U.S. when Vice President Al Gore called him on December 11, 1997, from Kyoto. In response to his questions, Gore told him that the nations of the world favored signing the Kyoto Protocol, whereupon Clinton said to him to go along and sign for the U.S. as well. And Gore did. But the decision had a sting in its tail. Clinton said that before Gore descended from his plane from Kyoto, the U.S. Congress had decided against the Protocol. Clinton said that the Kyoto Protocol had the historical distinction of effectively unifying the bitterly divided nation and the U.S. Congress, which was an important political goal of his. But this unity was achieved at his own expense, against him and against his decision in favor of the Protocol. Notably, the U.S. Congress at the time voted 95 to 0 against the Kyoto Protocol, and moreover against any treaty that did not commit developing countries to "meaningful" cuts in emissions.

The Protocol is an extraordinary outcome for the EU and the U.S., since it accepts the targets proposed by the EU, but the overall structure of the Protocol and its carbon market came from the U.S. The Kyoto Protocol is an American invention. Indeed, the overall structure follows my market strategy, and in this sense it follows the U.S.'s own market-oriented position, modified by a more favorable treatment of the developing nations in terms of no emissions limits and the addition of the CDM. The CDM allows industrial nations to receive credits for proven emissions reduction projects that are carried out in developing nations. These credits can be traded in the emissions market, so they carry all the advantages of the trading system without requiring emissions limits, just "baselines" on developing nations.

The Kyoto Protocol has a flexible and market-oriented architecture. The structure of the Kyoto Protocol is an agreement on emissions limits country-by-country, a great achievement for its lead negotiator, Raul Estrada Oyuela, with three flexibility mechanisms to accommodate those nations that in a given year may be above their limits, while maintaining fixed global limits: (i) joint implementation, (ii) the carbon market and (iii) the CDM. The most important and innovative feature of the Kyoto Protocol is the carbon market which, together with the CDM, achieves the level of equity that is needed to achieve market efficiency in using the planet's atmosphere.

The developing nations do not trade in the carbon market because they have no limits on emissions themselves. Only the participating OECD nations do. Yet the developing nations participate in the carbon market through the CDM, a strong incentive that allows businesses in rich countries to offset their emissions by funding clean energy projects in developing nations. In the Kyoto Protocol, the CDM is the only link between the industrial and the developing nations. It is the best hope for the future. It creates incentives for developing nations to adopt clean technologies and 'leapfrog' into the future, using a cleaner form of development than the industrial nations used to develop themselves. Ambassador Oyuela achieves his mandate of reaching an agreement in Kyoto acting against his own convictions. He does so — as a true professional diplomat — using the carbon market as the two-sided coin that resolves the conflict between the North and the South, and against his own anti-carbon market opinion. The carbon market helps him reach an agreement, and the Kyoto Protocol is born.

Post-Kyoto

In 2001, the newly elected U.S. President George W. Bush denounces the Kyoto Protocol saying that it will damage the U.S. economy. To date the U.S. has not ratified the Protocol. That same year, the third IPCC Assessment Report declares that the evidence that global warming over the previous 50 years had been fuelled by human activities is stronger than ever. The Kyoto rules are finalized in 2001 at COP7 in Marrakech. The Marrakech Accords provide no quantitative limits on emissions trading, significant credits (removal units) for forest and cropland management, caps on CDM, credits for sink activities and no credits for avoided deforestation. The current situation is of growing scientific concern; the evidence continues to reinforce the genuine threat of global warming, and only a handful of outliers now dispute these findings. Although Kyoto enters into force without the U.S. and without limits on developing nations, it may not be sufficient to prevent climate change.

In 2003, Europe experiences one of the hottest summers on record, causing widespread drought and heat waves. As a direct result 30,000 people die.

In 2005, following ratification by Russia in November 2004, the Kyoto Protocol becomes a legally binding international treaty. The U.S. and Australia continue their refusal to sign up, claiming that reducing emissions would damage their economies.

By 2007, 175 countries have ratified the Kyoto Treaty. Under its newly elected Prime Minister Kevin Rudd, who ran a campaign based on changing Australia's policy towards the Kyoto Protocol, Australia, too, ratifies the treaty. The IPCC Report for a fourth time states that "warming of the climate is unequivocal," and that the levels of temperature and sea rise in the 21st century will depend on the extent or limit of emissions in the coming years. Former Vice President

Al Gore and the IPCC jointly win the Nobel Peace Prize for services to the global environment. The Kyoto Protocol is an international law and without time limits; however, the emissions limits in its Appendix have two commitment periods, the first in 2015, now expired, and the second is in 2020.

At an earlier Convention of the Parties in Buenos Aires, the U.S. was completely unwilling to discuss the post-2012 period. And it was joined in this position by important developing nations such as India. Yet in Bali on December 2007, the Convention of the Parties of the UNFCCC decides on a so-called Bali Road Map, to arrive at the terms for a post-2012 agreement by the end of 2009. A great step forward is achieved when the largest emitter in the world, the U.S., agrees to join this effort by the 2009 target. This is the first sign of U.S. cooperation with the Kyoto Process since the U.S. signed the Protocol in Kyoto on December 11, 1997.

In 2008, Australia becomes the first nation to create its own internal carbon market, to start trading in 2010. Meanwhile, 414 sq. km (160 sq. mi) of the Wilkins Ice Shelf breaks away from the Antarctic coast. A year later, in 2009, an ice bridge connecting the Wilkins Ice Shelf to the Charcot Island breaks, leaving the ice shelf vulnerable to further collapse.[17] Scientists become concerned that climate change may be happening faster than previously thought. Today, the Australian Great Barrier Reef is half gone, as is the coral in the Caribbean Sea.

Following the Bali Road Map, negotiators from 180 countries launch formal negotiations toward a new treaty to mitigate climate change at the Bangkok Climate Change Talks in April 2008. At this meeting the EU announces their desire to cap or otherwise reduce the CDM transfers to developing nations, which amounted to about $9 billion until 2007 and to $18 billion in 2007, arguing that Europe needs to spur new technologies because simply paying for off-takes elsewhere will not solve the problem. The CDM advocates increasing payments to clean technology, meaning $100 billion in total. By 2011, the carbon market records trading of $176 billion annually. Yet business interests are now working to preserve and expand the CDM program as the EU proposals move through the European Parliament. At the time of this writing, Ambassador Raul Estrada Oyuela is still opposed to the concept of the carbon market as part of the global climate negotiations.

Here are other important milestones in the international climate negotiations

1988: IPCC establishes the scientific basis of climate negotiations.

1992: UNFCCC is established.

1995: COP1 in Berlin establishes the goal of reaching a climate agreement by 1997.

1995: The first definitive statement that climate change is caused by human action is released.

1997: COP3 in Kyoto — The Kyoto Protocol and the carbon market are born.

2001: The Marrakech Conference establishes the Kyoto rules.

2005: The Kyoto Protocol and its carbon market become international law.

2007: The Bali Road Map is established for reaching a post-Kyoto agreement by 2009.

2009: Copenhagen COP15 is announced with great hopes for a global accord, but ends in disappointment.

Yet, at the COP15, I introduce the concept of carbon negative technology to the CDM, then chaired by Christiana Figueres, now enshrined as the only possible solution in the IPCC Fifth Assessment Report; create the Green Power Fund (GPF) and publish a sequence of articles in the *Financial Times*. My proposal for the GPF submitted to the U.S. Department of State, is accepted partially by then Secretary of State Hilary Clinton and later voted on as the Green Climate Fund in Durban South Africa COP17.

2011: Durban South Africa COP17 extends the end of the Kyoto carbon emissions limits from 2012 to 2015. It approves the Green Climate Fund instead of the Green Power Fund without connection to the carbon market nor CDM, and with no agreed sources of funding nor uses for the funding except for mitigation and adaptation to climate change.

2012: The initial Kyoto Protocol Carbon emissions limits provisions. The Doha Amendment extends the Kyoto Protocol until 2020.

2013: The IPCC Fifth Assessment Report for the first time mentions that "carbon removals" (carbon negative technologies) are required to avert catastrophic climate change.

2014: COP20 in Lima, Peru Call for Climate Action announces COP21 Mandate.

2015: COP21 in Paris seeks global accord with universal participation and technology solutions for climate change. No mandatory emissions limits achieved.

2017: The U.S. announces its withdrawal from the Paris Agreement.

2020: New universal regime starts. Protocol developed by the Ad Hoc Working Group on the Durban Platform for Enhanced Action (ADP) comes into effect.

Behind the institutional facade, the parties of the climate negotiations have followed predictable patterns of behavior, most of which continue to this date. It is useful to understand these patterns because they explain where we are today, how we got here and what can and should be done for the future of the climate negotiations. The future of the climate negotiations seems as uncertain as climate evolution. Perhaps even more pressing is the Kyoto Protocol's provisions for emission limits expiring in 2020, putting a big question mark over the future. COP21 was a dividing line and there were no clear new achievements. The Paris Agreement was an expression of hope but nothing concrete has been achieved since 2015.

We can, and must, expand on Kyoto's initial provisions. But to forge global agreement, we must identify how to reconcile the needs of developed and developing countries. Kyoto's innovative carbon market has been an important measure of success and provides the mechanism to achieve this. In 2015, the world's six largest oil and gas companies openly wrote to the U.N. asking for more action to create a "price on carbon" which the carbon market provides.[18]

The carbon market is a path forward as it provides funding for the CDM, the Clean Development Mechanism. Now the CDM requires improvement to provide funding for carbon negative technologies.

The next chapters explain the economic changes unleashed by the carbon market, the hopes of the Paris Agreement and its failures, and a path to the future of the negotiations. The goal is a resolution of the climate change crisis.

5

An Uncertain Future

Bali is one of the most beautiful places on earth, a large idyllic island in the Indonesian archipelago, home to historic art and mischievous monkeys.

But it is also one of the places most vulnerable to rising sea levels from climate change. This was a poignant reminder to climate negotiators of just how much was at stake when they met at the Bali Convention of the Parties (COP) in December 2007.

The focus of the Bali meetings was what to do after 2012, when the Kyoto Protocol emissions limits were set to expire. The final date was now shifted to 2020 in Doha, Qatar, with the adoption of the "Doha Amendment to the Kyoto Protocol."[1]

International negotiations over how to achieve emissions reductions beyond 2012 began in 2005, when COP11 met in Montreal to mark the entry into force of the Kyoto Protocol, when it became international law. But the celebration did not last long.

By the time of the COP13 meeting in Bali there was new scientific evidence suggesting that global warming was happening even faster than we had feared. The gap between Kyoto's proposed emissions reduction and the rapid global increase in carbon emissions had increased. This provided a renewed sense of urgency to the climate negotiations.

The COP13 meetings in Bali culminated in the Bali Road Map, which launched the new negotiation process to determine Kyoto's future. The roadmap ensured at least two more years of talks before the process was to be completed at the COP15 meeting in 2009 in Copenhagen. The process would determine for years to come how the world would reduce the emissions of greenhouse gases. The fate of the planet was supposed to be determined at Copenhagen in 2009. Time would tell.

An Insider's Account of the Bali Negotiations

For me, Bali had the usual circus atmosphere of the United Nations Climate Convention events, this time in an exquisite setting. I was invited to participate in a side event by Hernan Carlino, the Chairman of the Clean Development Accreditation Committee and the representative of the group of developing nations (G77) in the Executive Committee of the Kyoto Protocol. My role was to present a proposal for a follow-up to the Kyoto Protocol after 2012.

The weather was warm and clear. The debates were, as usual, highly technical and, this time, the meetings were televised and viewed by the millions of people around the world who care about the future of Kyoto.

Yet by all accounts, the frustration of most negotiators at the meetings was palpable. At a time when most countries were ready to make a firm agreement about the next steps, the U.S. delegation insisted that more debate was needed. The tension finally erupted in a David and Goliath-style showdown when the representative from Papua New Guinea,

Kevin Conrad, boldly proclaimed to the U.S. in a televised statement that sent shockwaves around the world, "…I would ask the United States, we ask for your leadership, we seek your leadership, but if for some reason you are not comfortable leading, leave it to the rest of us, please get out."[2]

Papua New Guinea is a small traditional nation with a large agenda: the preservation of their forests and their biodiversity. Despite appearances to the contrary, this nation is actually rather favorable and friendly towards the U.S. Until this point in the negotiations at Bali, the U.S., as usual, resisted, and was less than cooperative. After so many years of U.S. hostility to the negotiations, many nations naturally wished that the U.S. representatives were not there in the first place, so as to allow others to move on with the agenda. Not so with Papua New Guinea, who was friendly to the U.S. After its bold statement, which was actually intended to bring the U.S. back into the fold, the U.S. made a surprising about-face and agreed to join the initiative to negotiate a follow-up agreement to Kyoto — the Bali Road Map. To the world's surprise, the government of U.S. President George W. Bush, one of the most resistant and hostile administrations to environmental concerns in U.S. history, agreed to follow the Bali Road Map, with the aim of joining the Kyoto Process at the end of 2009. This was a historic moment, misunderstood perhaps, but full of hope for change. Time would tell.

The meetings in Bali and the Bali Road Map launched a new two-year negotiation process to tackle climate change. Like the 1996 Berlin Mandate that paved the way for the Kyoto Protocol, the Bali Road Map was an agreement to agree. It ensured that the global community would embark on at least two more years of negotiation towards fashioning a successor to the Kyoto Protocol. The process would be completed at the COP15 meeting in Copenhagen at the end of 2009.

Kyoto's future had never been more uncertain. At the April 2008 Bangkok meetings, discussed in Chapter 7, the EU threw a monkey wrench into the negotiations that marked the start of the Bali Road Map. A failure of the Bali process could effectively terminate Kyoto. This revealed the uncertainty of the Protocol's future.

We had barely set out on the road and already negotiations had stalled. Helping poor nations chart a new clean path to industrialization is central to the climate talks. But the atmosphere of the climate negotiations was not good. The conflict we faced was clear: it was once again a global conflict between the rich and the poor nations, but in my view, this conflict was created by a lack of understanding of what Kyoto could achieve and how. In reality it could unify the interests of both.

Bali COP13 was a larger meeting than ever. Non-governmental organizations (NGOs) made up the majority of the participants. At the meetings, many dedicated environmentalists carried signs with dire predictions about the future of humankind and offered good omens to Kyoto. Dr. Robert Watson, the former Chairman of the Intergovernmental Panel on Climate Change (IPCC), was there, as was Professor Robert Stavins of Harvard University, who organized a well-advertised and well-attended side event on "Kyoto after 2012." A side event is an official presentation by outsiders. This event had the same topic as my own presentation in Bali, eliciting the participants' proposals on what to do about Kyoto. But in this case, no proposals were made. In response to a question he posed to the audience, I asked for his own view of what to do, but he offered no response.

New Proposals for the Post-Kyoto Regime

In this context, I presented my proposals at a panel together with Professor Peter Eisenberger of Columbia University and

Professor Hernan Carlino of the University of Buenos Aires and the United Nations Framework Convention on Climate Change (UNFCCC). The panel was not well advertised and there were not many people in the room, but the participants were keen and intense and the presentations were officially recorded for posterity. Our presentations highlighted a proposal for the post-2012 Kyoto regime in two parts. The first part was a new type of negative carbon technology that we proposed to be incorporated as part of the Kyoto Protocol's Clean Development Mechanism (CDM). We argued that the technology, which is based on capturing carbon from air, was needed to prevent global warming in a timely fashion, and could help poor nations; by capturing more carbon than they emit, these nations could earn credits in the carbon market. The second part of the proposal involved a global financial mechanism that builds on the carbon market itself and could potentially resolve the impasse between developed and developing nations, which has been called the China–U.S. impasse.[3]

The two proposals could resolve the core of the conflict between poor and rich nations, and the most important obstacles that Kyoto faces today.

The technology we proposed reduces atmospheric carbon while producing electricity. It uses the same energy for both processes, so it "cogenerates" carbon reduction and electricity production. This unusual technology can greatly accelerate the reduction of carbon from the planet's atmosphere as required by IPCC targets, while at the same time increasing the production of energy in the world. This one–two punch of more energy and less carbon seems an almost impossible combination to many people. It appears almost too good to be true. Yet it is, in reality, the latest generation of a tried-and-tested technology called Carbon Capture and Sequestration (CCS), which has been in operation since 1996, and was recently the subject of a McKinsey & Company report in the EU.[4]

CCS methods have been used for over 13 years by the oil industry to "scrub" carbon from its production processes, and inject it into oil deposits to enhance oil recovery, which it does by as much as 30–40%.[5] The following sites have been operating CCS for several years, and each captures about 1 million tons of carbon dioxide a year: Sleipner, Norway, (offshore) since 1996; Weyburn, Canada, since 2000; Salah, Algeria, since 2004 and Snovit, Norway, since 2008. The best that can be achieved with this conventional CCS process is "carbon neutrality," namely to clean up all the carbon that the plant emits, and no more. However, in Bali, I proposed a different approach, a new generation of CCS that is "carbon negative" rather than "carbon neutral," by extracting carbon directly from the atmosphere, and trademarked the term "carbon negative technologies™". The American Physics Society (APS) was then writing a report on this new technology.[6] The negative carbon approach is drastically different and more effective than the carbon neutral method because it actually reduces the concentration of carbon in the atmosphere and can do this even in the process of producing electricity.

The properties of this new technology were surprising to many and Carlino thought it would be important to bring a discussion of them forward in the context of the Bali COP, when the future of the Kyoto Protocol would be discussed. We had discussed the possibility of the technology a year earlier in August 2006 in Buenos Aires, Argentina, where I was giving several presentations and touring Patagonia, with its spectacular whales and its huge — now melting — glaciers. We decided at the time that negative carbon technologies, despite obstacles to implementation and innovation, could benefit from the funding provided by the Kyoto Protocol CDM for clean energy. For this, we only needed to modify the CDM to accept it. The technology is best suited to developing nations that do not emit much carbon, such as African or Latin American nations because negative carbon allows them to capture more carbon than they emit, which is little. This way these nations could sell a substantial amount of carbon credits in the carbon market. For example, Africa currently emits about 3% of global emissions and, under current CDM practices, has little carbon emissions to reduce and, therefore, little to sell in terms of carbon credits. For this reason, by 2013 most of the CDM funding for clean projects (involving the registration of 7,870 projects and programs in 107 developing countries with an estimated $215.4 billion in total investments) went to the developing nations that emit most carbon, such as China (see Figure 1 "Distribution of Issues Certified Emission Reductions by Host Party" for the breakdown).[7]

Using negative carbon processes, such as carbon capture from air, algae's capture of carbon, etc., Africa could capture, for example, 20% of world emissions, even though it only emits about 3%. The carbon it captures could be sold in the carbon market, benefitting African nations commercially as well as reducing the world's carbon concentration in the atmosphere. Private sector firms are adopting carbon prices, and internal carbon pricing is now used by at least 150 companies as reported by the Carbon Disclosure Project (CDP), with prices from $6–89 per ton of CO_2.[8] In

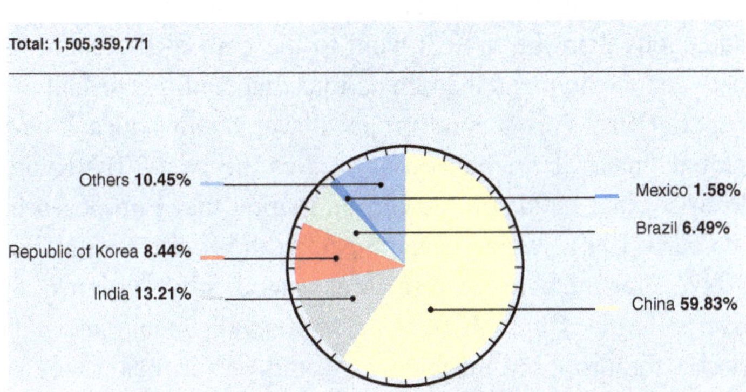

Figure 1 Distribution of Issues Certified Emission Reductions by Host Party

addition, if the negative carbon technologies can augment Africa's energy production — by cogenerating electricity and carbon capture as mentioned — then there is a triple benefit for Africa and for the world: Africa can develop more energy plants and sell more carbon credits in the carbon markets, enhancing African development, and the rest of the world can benefit from lower carbon levels in the atmosphere as well as from, in the case of the Organization for Economic Co-operation and Development (OECD) nations, increasing exports of technology to Africa for its new cogeneration plants. It is a truly remarkable and realistic possibility that the CDM reached $13.9 billion in annual investment for operating projects, since the total for all projects was $40.4 billion in 2008, and the number of registered projects kept growing. Indeed, the CDM could do a world of good for developing nations.[9] The inexpensive and clean energy generated in these negative carbon projects would allow development in Africa and Latin America as nothing has done before. Development requires energy. Reciprocally, by making the Kyoto targets easier to achieve, this technology could make Kyoto targets more realistic. This, in turn, supports the Protocol's continuation into the future. It is a win–win approach.

My second proposal at the Bali event was even more audacious than the first. It went to the core of the conflicts between the poor and the rich nations that continue to stall the negotiations. This second proposal was to introduce a new global financial mechanism as part of the post-2012 Kyoto Protocol that builds on the carbon market that I proposed in the early 1990s, which became part of the Kyoto Protocol in 1997. This new mechanism was specifically designed to overcome the China–U.S. impasse. For all intents and purposes, the proposed mechanism would perform as a way to introduce limits on Chinese and other developing nations' emissions without contradicting Article 4 of the U.N. Climate

Convention: it could provide assurances to the U.S. and would at the same time compensate China for carbon reductions. The proposed financial mechanism would replicate in financial terms almost exactly the wording of Article 4 of the 1992 Climate Convention of the United Nations (in Chapter 4). I had presented this proposal at the International Monetary Fund in Washington D.C. twice, in August and in October 2007. On each occasion, the audience responded to the proposal favorably, but with somewhat startled reactions. My official presentation at the Bali COP, in December 2007, was received with the same combination of interest and surprise.

In the ensuing months, Carlino would leave his position at the CDM UNFCCC Committee. Yet the discussions continued, and I presented the proposals on November 12, 2008 to the Australian Parliament in Victoria at a briefing organized by MP Michael Crutchfield, receiving an enthusiastic response from the parliamentarians and from prominent Australian businessmen. This led to a request for another briefing with their Minister of the Environment, Gavin Jennings, and with their Ministry of Primary Industries and its Deputy Secretary, Dale Seymour, for the purpose of creating a demonstration plant for this technology. The site was Gippsland Victoria, home of the largest brown coal deposits in the world.

"Et tu Brute?"

The climate negotiation process is long and arduous, and only toward the end does it becomes clear whether a particular event is a failure or a success. The Bali Road Map process developed throughout 2009, and only led to conclusive results in November and December 2009. But it still had significant obstacles to overcome.

The Kyoto Protocol was an important first step toward the UNFCCC goal of a low carbon global economy that provides for *basic needs* while respecting ecological limits. Yet, despite having just started, the Kyoto Protocol seemed to already be on its last legs, just a few years after it came into force. This time, the threat came from an unlikely enemy. Recent European proposals, supported by Japan, could destroy what Kyoto had already accomplished and its promise for the future. The EU and Japan had been the strongest and most loyal supporters of global climate negotiations so far, so why were they suddenly changing their course?

The challenge was an old threat with a new face. In April 2008, European negotiators went to a United Nations climate meeting in Bangkok. They warned the representatives of developing countries that they needed to step up to the challenge of climate change if they were to see additional money flowing into CDM projects in their countries. Their words seemed reasonable; we all needed to step up to the challenge of climate change. But the motivation behind the words was alarming. What the EU proposed threatened the one policy that could most effectively solve the challenge of climate change.

The EU proposed to cap CDM investments, which are profitable for industrial nations and create clean technologies in developing nations. These investments prepared the world for a future with fewer emissions. The process begins with nations that are projected to be the largest emitters in the future, even though they may not be the largest emitters today. If successful in limiting the CDM, the EU would succeed in dismantling an important tool for future emissions reductions.

The recommendations in Bangkok reflected a misunderstanding of what Kyoto intended, and of what, in fact, Kyoto had already achieved. The recommendations by the EU were intended to boost emissions reductions at home in Europe, while giving "developing nations incentives to do more than sit back and watch the money flow."[10] But increasing investment in clean technologies for developing nations is an important part of what the CDM has achieved thus far. The purpose of the CDM goes beyond just transferring money to developing nations — although this could be considered a worthwhile goal. The purpose of the CDM is to create hard-earned and profitable projects that transform development in the poorest nations into a new kind of clean development. By design, the CDM is intended to enable developing countries to move into a new type of industrialization that the entire world badly needs.

There is increasing recognition that industrialized nations, who are historically responsible for climate change, cannot adequately address the climate crisis alone, given the rapid rise in emissions in emerging economies. But in a somewhat perverse response to the problem, the European Commission proposed a set of policy proposals that would scale back the CDM, the only mechanism under the Kyoto Protocol that provides incentives for reducing emissions in developing nations. The EU proposals would essentially cap the CDM at current levels until 2020 if a new climate treaty were not reached. Commission officials say that a moderate expansion would be allowed were an agreement in Copenhagen reached, although estimates vary as to the actual impact.

Looking beyond the strategic errors and misunderstandings, the European Commission had an important point. The CDM program was in need of reform. Indeed, it may not be the right model to solve many problems unless and until it is reformed, which could be achieved with modest changes. For instance, developers were now requesting new CDM credits for hydroelectric, wind power and, most recently, natural gas-fired power plants in China. There is some perversity in the CDM program if 60% of all CDM projects fund changes in China's energy structure, while the poorest nations in the world are left out because they emit so little and therefore cannot reduce their emissions significantly.

CDM projects, when they are carefully monitored, can guarantee that off-takes represent verifiable emissions reductions above and beyond business-as-usual baselines. This monitoring is done by the UNFCCC Kyoto Protocol Accreditation Committee, but it is a bureaucracy that requires streamlining to make it possible for small and poor nations, such as Bolivia and Mongolia, with little or no investment banking know-how, to compete for CDM projects.

Despite these problems with the CDM, problems that can be successfully addressed, the logic of the new EU proposal was flawed. It was so flawed that it had created strange bedfellows — the poorest nations of the world and the rich nations' businesses. Poor countries feared that the proposal would dry up the steady flow of funding for investment in clean technologies that had been available since 2005. Businesses feared that the recent proposals would throw the entire carbon-offset industry, then valued at $50 billion, into flux.[11] "The cap on clean development projects proposed by the EU, as it is designed now, will not provide any incentive for people to design new (clean technology) projects. Effectively the market will be killed." So spoke Michaela Beltracchi, European Policy Coordinator for the International Emissions Trading Association, based in Geneva, Switzerland, which represents a range of business interests.[12]

European nations had not done enough to ready themselves for a carbon-neutral future. They would fail to achieve their committed Kyoto targets by 2012. Italy, for example, was registering emissions growth of 13%, when it is supposed to meet an emissions cap of 6.5%.[13] But the CDM was not to blame. One widely acknowledged problem was that the EU set too lenient limits on carbon emissions as part of its own internal cap-and-trade program. This depressed the price of carbon permits in the EU carbon market, provided less incentive for industry to invest in emission reductions within the EU. Their experience is now hailed as an example of what not to do with cap-and-trade. In the U.S. the emerging consensus is that carbon allowances must be sold, perhaps in public auctions, to emitters.

The Present vs. the Future

In the ensuing debate there were two powerful opposing points of view that we already discussed. One represented the past and the other represented the future. Advocates of the Kyoto Protocol represented the future. Critics argue that most of the emissions reduction projects in the CDM would have proceeded anyway, and so do not represent additional reductions. Along the same lines, critics have argued that the CDM emissions reduction may not be credible. Those concerns must be taken seriously.[14]

The EU believes that Europe itself needs to advance technology, and that paying for off-takes elsewhere does not prompt the technological innovation the EU needs to solve its own problem. Others believe that pulling the plug on the CDM will have little effect on the climate talks. They believe that the CDM could go and China would not even notice.[15] All of the CDM projects in China, which account for over one-third of the entire CDM project base,[16] represent less than 1% of the country's annual growth.

In a startling about-face, business interests started working to preserve and expand the CDM. This was happening as the EU Commission's proposal against the CDM moved through the European Parliament that year. One proposal was to leave the overall cap on European emissions in place and to relax international caps in all nations. This would ensure that investments in new technologies are made at home, since the overall cap on European emissions remains the same, while allowing the CDM to grow and expand.

Another big question for the future was how international CDM credits would be handled by the U.S., whose carbon market could dwarf the EU's. The leading climate negotiations in the U.S. Senate could limit international offtakes to 15% of the overall market. Critics suggest that the U.S. legislation could unwillingly incentivize other nations to "launder," in other words illegally sell, international credits with no oversight from the U.S., allowing the global emissions to go above what was intended.

But the U.S. itself was involved in another major global conflict: the competition for natural resources by developed nations, like the U.S., and developing nations, like China. This conflict manifested itself in the U.S.'s refusal to ratify the Kyoto Protocol or to participate in the carbon market altogether unless China would accept limits on its own emissions. The uncertain future of the Kyoto Protocol is perpetuated by the China–U.S. Impasse, which is truly dramatic. This is global politics occurring on the world stage: the competition between an old superpower and a new superpower. It is the oldest type of competition there is, and the world's climate is hostage to its resolution.

The issue exists in the realm of global geopolitics at the largest possible scale. It is the new geopolitics of the 21st century. Rich and poor nations are no longer competing for nuclear power as they did during the Cold War era, although the risks and rewards of nuclear power have not gone away.

They are competing for natural resources and, in the case of global climate change, for the rights to use of the atmospheric commons.

Yet the nations of the world have a new terrifying and unifying risk to contend with: global warming. The similarity with nuclear risks is not idle. As in the case of nuclear power, the risks we face can unify us, precisely because they are so large and potentially catastrophic; the survival of human societies may be at stake. This is an unexpected silver lining in the global warming cloud: global warming has a capability for unifying the human race like nothing has done before.

For the first time in history we are all in the same boat. For the first time, poor nations are in a position to have a major impact on the standards of living, even the survival, of rich nations. Simply by expanding their economies through their own energy resources such as coal and petroleum, African nations have the ability to incur trillions of dollars in losses for the U.S. The OECD reports that Miami is at risk of $3.5 trillion in property damages if African countries induce a rapid increase in the sea level, simply by burning their own coal and oil.[17] Until now, a U.S. citizen could be concerned about the standard of living of African people but not have a direct self-interest in the matter of how African nations grow their economies. This is certainly the first time in history that we can say that African countries have the power to considerably reduce the standard of living in the U.S. Although this sounds like a grim outcome, it could induce positive results. Now the U.S. has a real, *bona fide* self-interest in the clean development of African nations. And the same is true about the clean development of Latin America. These are the two geographical areas from which the U.S. imports most of its fossil fuels and natural resources.

Rich and Poor: Who Should Abate Emissions?

We all know that developing nations hold the key to the future. They will become the largest emitters as they industrialize. In 1997, on a per capita basis, industrialized countries in Annex I of the UNFCCC had emissions that were, on average, 2.5 times more than those of developing countries. Yet, since the IPCC Fourth Assessment Report in 2007, total emissions from countries not listed in Annex I have overtaken the emissions of the Annex I developed countries.[18] To avoid worsening this trend, developing nations should follow a clean path to industrialization, something that is obviously not happening. In 2014, China built one out of every two of the world's new coal power plants every week.[19] China is today the world's largest emitter of CO_2.

The issue is how to ensure a clean future for the poor nations of today. One question that was widely discussed is whether or not to cap developing nations' emissions. This is specifically forbidden in Article 4 of the 1992 UNFCCC, which states that unless developing nations are compensated they will not be required to reduce emissions. The issue is so controversial that it can be called the "third rail" in international climate debates. If the issue is not resolved, negotiations could break down without an agreement for what should follow the Kyoto Protocol. The U.S. has already stated its unwillingness to accept limits on its greenhouse gas emissions unless China accepts the same. But now they appear to be uniting toward the EU's call to cap developing countries' emissions. The EU had said that developing nations should accept caps of 15–30% of their business as usual emissions.[20]

First, the nations have to agree to lower the global cap on emissions. The Kyoto Protocol mandates an average emissions reduction of 5.2% of 1990 levels. So far, this excludes the U.S., who did not ratify the Kyoto Protocol, so the

Protocol accounts for only part of the emissions from the industrialized world. This was a good start, but it was not enough. Further emissions reductions need to take place after the first commitment period. To avoid major climate disruption the IPCC urges an 80% cut in emissions within 20 to 30 years. The international community understands the need for a lower emissions cap; what it does not agree on is how to reach it. Do we increase the caps for developed countries? Do we impose caps on developing countries?

Warming Up to Climate Action

One thing is clear: we cannot significantly reduce emissions without U.S. participation. On this front, things are not nearly as bad as they were before. The U.S. is the world's second largest emitter — about 25% of the world's emissions come from the U.S. — and it has been the strongest opponent of the Kyoto Protocol. Things were changing; hundreds of U.S. cities and towns have signed petitions demanding that the federal government in Washington ratify the Kyoto Protocol and join the carbon market. In 2007, the U.S. Supreme Court agreed that the U.S. Clean Air Act of 1963 gives the federal government the authority to regulate and limit greenhouse gas emissions. Still, under the Trump administration, the U.S. withdrew from the Paris Agreement.

Since the U.S. abandoned the Kyoto Protocol in 2001, climate policy had slowly but surely advanced at regional, state, and local levels within the U.S. The Regional Greenhouse Gas Initiative (RGGI), a collaborative effort of 10 northeastern and mid-Atlantic states, had mandated a 10% cap on carbon emissions from the power sector by 2018. This is the first mandatory, market-based effort in the United States to reduce greenhouse gas emissions. Permits will be auctioned and revenues will be invested in energy efficiency, renewable and other clean technology projects.

The Western Climate Initiative, which involved seven western states and four Canadian provinces, has been moving toward the creation of its own regional cap-and-trade program. These two regional initiatives could allow some U.S. states to join the carbon market. In addition, the first federal climate policy initiative, the Lieberman Warner Climate Security Act, reached Congress in the spring of 2008. It, too, was based on a cap-and-trade system for emissions reduction and although it was defeated across Republican and Democratic Party lines, it set the stage for further negotiation over climate policy.

Then-President Barack Obama acknowledged the urgency of the climate crisis and had called for national action to combat climate change. In particular, he announced his support for moving the U.S. back into the Kyoto Protocol. Environmental organizations across the U.S. were geared up for a national cap-and-trade program and expected ratification of the Kyoto Protocol. President Obama and the USEPA created the clean Power Act in 2014, which when implemented is the basis of a national carbon market; however the execution was delayed by legal challenges in several states.

Business interests in the U.S., the bastion of innovative capitalism, had warmed to climate policy as well. Silicon Valley Venture Capital is already investing a significant part of its risk capital in clean energy projects, and it is believed that this could increase rapidly. Analysts in major investment banks, such as J.P. Morgan Chase, now routinely use carbon foot-printing to evaluate a company's risk profile. Despite the hostility of the current administration, it is well known that more jobs have been created by the renewable energy industry than the entire oil and gas industry, in the U.S. and also globally. Business interests acknowledge that the U.S. automobile industry is handicapped from benefitting from the "price signal" of the carbon market that helps orient industry elsewhere to build vehicles with lower carbon emissions.

To Cap or Not to Cap: Is that the Question?

Actually, in the end, it does not matter. CO_2 levels are too high already. Arguing about who or what is responsible is useless. The only certainty is that the climate is changing. This is a war, and the battles are for survival. Discussions and conversations have lasted long enough. It is time to lead, to follow, or to get out of the way.

But let us examine the situation one last time.

The policy of rich nations investing toward poor nations is essential to the climate negotiations. It was the foundation of the CDM, which unifies the interests of businesses in the rich nations with the interests of poor nations in clean development. This community of interest between the North and the South, and the importance of the role of developing nations, is grounded in the Climate Convention of 1992. It is, in fact, memorialized in its Article 4. The idea is that developing nations should be able to increase emissions for a time to grow their economies and lift their citizens out of poverty. The World Bank reported that approximately 32% of the people in the world live on less than \$2 per day in 2011, and roughly 1.2 billion people live on the brink of survival, with incomes of less than \$1.25 per day.[22] The developing nations export most of the fossil fuels that are used in the world, but they neither use most of the fossil fuels nor produce most of the world's emissions. Latin America and Africa are the main resource exporters in the world economy. They have exported their resources at prices that are so low that poverty has taken grip of their people and the rich nations have become "addicted" to their fossil fuels and other resources.

Many poor nations hardly consume energy at all. Consequently, they produce very few emissions; in contrast, all the industrial nations together produce about half of the world's carbon. Poor nations are not the main cause of the global warming problem. Nor can they be the solution. In reality, poor nations are the main victims of global warming risks. Over 80% of the planet lives in the developing world, where we will see the worst consequences of global warming: desertification, agricultural losses, interruption of water supplies, and terrifying exposure to rising sea levels.[23]

It made sense for the Kyoto Protocol to grant developing countries unlimited rights to the global atmospheric commons on equity grounds. But it was much easier to make this concession in 1997 than it is now. Energy consumption and emissions production in the developing world was so low that these countries offered little potential for emissions reductions. Because poor countries did not emit much in the first place, it was obvious that developed nations had to shoulder the burden of global emissions reduction. But this is under consideration, since China is now the largest emitter, despite being a developing nation.[24]

The good news is that developing nations are growing in economic terms — some much faster than others, and none with any guarantee that growth will trickle down to their poorest, but growing nonetheless. The bad news is that as developing nations are growing, they are consuming more energy in the process. There is a clear and direct connection between energy use and economic output. A country's industrial production can be measured from its use of energy.[25] Emissions from developing nations are increasing at a growing rate. The emissions growth rate in the developing world is higher, on average, than it is for all other countries, including the U.S. In the U.S., carbon dioxide emissions are projected to increase at an average annual rate of 1.1% from 2004 to 2030. Emissions from developing nation economies are projected to grow by 2.6% per year.[26] As a result, developing countries' share of global emissions will rise. China now has surpassed the U.S. as the world's largest emitter, producing 10,975.5 MtCO2e in 2012 compared to 6,235.1 MtCO2e in the U.S.[27] This is all very persuasive evidence for capping emissions growth in developing nations. But there are equally persuasive arguments for not imposing emissions caps on low-income countries. Global income inequality is more acute now than it has ever been in human history. Inequality between nations is larger than inequality found within any single country, including Brazil, South Africa, and the U.S., where income inequality is known to be high. The top 5% of people in the world receive about one-third of total world income. The top 10% of people in the world receive one-half of the world's income. The ratio between the average income of the richest 5% and the poorest 5% of the world is 165 to 1. Roughly 70% of global inequality can be explained by differences among countries, rather than within.[28] To tackle the global divide, we need to increase income growth in the poorest nations. This is much harder to do in a carbon-constrained world unless we can develop and transfer technologies to the developing world for renewable and clean energy resources and greater energy efficiency. Additionally, we must create incentives to reduce emissions in the developed world.

What about China?

Should China be treated differently? It is tempting to treat China, and India to some extent, as distinct from the rest of the developing world when it comes to global climate. Indeed, there are some important differences. China's annual economic growth rate trumps the growth rate of all other economies worldwide. China's economy is now the size of the U.S. economy. But with a population of over 1.3 billion people, most of whom live in China's impoverished rural areas, per capita income in China still is about $2,360. In the U.S., per capita income exceeds $54,629.5.[29] Disparities this large still warrant attention.

China adopted massive energy policy reforms over the last two decades aimed at increasing energy efficiency and conservation. Between 1997 and 2000, China increased its emissions by 6.7%, while its economy grew by 26%.[30] China's sweeping measures represent emissions savings nearly equivalent to the entire U.S. transportation sector.[31] It is not only China that has taken large voluntary steps to reduce or slow the growth of their emissions. For example, both Indonesia and China are phasing out fossil fuel subsidies. China, Mexico, Thailand, and the Philippines have established national goals for renewable energy use and energy efficiency. Argentina and India are converting automobiles and public transport to natural gas.[32] The Costa Rican government recently announced its goal of making the country carbon neutral over the next 10 to 15 years.

In comparison, U.S. initiatives aimed at reducing greenhouse gases have been mostly voluntary and poorly coordinated across the country. In contrast, Chinese fuel efficiency standards are more stringent, as they are based on European standards. It is hard to expect developing countries to implement binding limits on their greenhouse gas emissions when the wealthiest of the industrialized nations is unwilling to follow suit.

Yet at the same time, recent news coming from China is a source of concern. Although it can still boast much lower per capita emissions than the rich nations of the world, the Chinese Academy of Sciences now reports that China's aggregate emissions tower above all other countries in the world, including the U.S., much sooner than anyone had anticipated.[33] This news increases the pressure on China to accept caps. It also indicates the enormous challenge China will face meeting any emissions target. Acknowledging the need to cap China's emissions is only the first step, and taking that step will be an uphill battle. But demonstrating the capacity of China to meet such a cap would be even harder. This is where the carbon market can help; it can channel investment into emissions reduction where the world needs it most — in China. China accepts this logic and has now adopted a carbon market.

The conflict between the U.S. and China has plagued climate negotiations since their beginnings. More generally, the conflict between the rich and the poor nations is the cause of Kyoto's growing pains. Why?

What is at stake in the global negotiations is fundamentally a question of who has the right to use the world's resources today and in the future, the rich or the poor nations.

How can poor countries pursue their economic development without jeopardizing the future for us all? If we perfect and expand Kyoto's global carbon market, it will create the potential for a new form of clean industrialization.

The global carbon market, if carefully regulated, could ensure an appropriate global cap on emissions and transparent trading. The result could change the world.

Once this is achieved, the CDM can provide powerful incentives to developing countries to reduce emissions, without resorting to caps. It effectively imposes a price on their carbon emissions, a price that they still have the flexibility not to pay, but at a cost.

Even without caps, developing countries could gain more from limiting emissions and selling credits through the CDM. In effect this means that developing countries still incur a "price" for their emissions. Every ton of emissions they produce represents a ton of emissions reduction they could have sold at the prevailing world price for carbon reduction. Economists have a special name for this: it is called opportunity cost. The opportunity cost of generating emissions when you could have sold emissions reduction credits to industrialized countries is in itself a powerful incentive not to emit.

This is a different incentive structure than what existed for all nations prior to the Kyoto Protocol, a time when all emissions were essentially unpriced and unaccounted for in decision making. Thanks to the CDM, developing countries face an incentive to reduce their emissions, even without caps.

And yet we still have not fully utilized the CDM, a potent feature of the Kyoto Protocol. The demand for buying emissions reduction credits through the CDM is less than its potential. The market is new and the learning curve for buying and selling in this market is steep. The cost of starting a CDM project is high and must be decreased. At present the costs are prohibitively high, in terms of both cash and know-how, for all but the largest projects. Time and improved streamlining of the costs of applying for CDM project certification can fix this part of the problem. But as long as the EU limits the use of the CDM, by maintaining that developed countries must meet most of their emissions reduction commitment "at home," the demand for these credits and the price developing countries can be paid for these credits will be less than its potential. And without emissions caps and a carbon market, there is no source of funding for the CDM. The 2015 Paris Agreement has no emissions caps.

We can increase the demand for CDM projects, and facilitate a greater transfer of technologies and capital to developing countries, by imposing more stringent caps on industrialized country emissions. To the extent that an expanded global carbon market can provide developed countries with greater flexibility, lower mitigation costs transfer technology and direct more energy for development, it makes it easier to convince industrialized countries to accept lower emissions caps for the future.

The carbon market saved the Kyoto Protocol once, by providing a mechanism for uniting the interests of poor and rich nations. We could perhaps do it all over again. Proposals on negative carbon and new financial mechanisms could save the day. This technology changes dramatically the perception of what is possible. That is what's required. As Einstein said, you cannot resolve a problem with the same mindset that produced it in the first place. Paris, however, proved to be a disappointment. And the 2017 COP meetings provided no solutions.

North–South: An Uncertain Future

International trade, particularly between rich and poor nations, has come, at many times, under intense scrutiny. Many acknowledge that international trade benefits rich countries more than poor countries.

Trade is often credited with increasing economic growth. A country's economic growth is measured by its Gross Domestic Product (GDP), a measure of the market value of all of the goods and services it produces over a given year. Yet goods and services are no longer a reliable measure of growth. Growth leads to more resource consumption and more use of the atmosphere. And GDP does not account for all of the harmful effects of our production and consumption. For example, the increased threat of climate risk associated with emissions created by industrial production is not reflected in fossil fuel prices or in the prices of goods and services produced with fossil fuels in the economy. Today, spending to clean up an environmental disaster, such as an oil spill, contributes positively toward GDP, since it creates demand for goods and services related to mitigation. For these reasons, as well as others, it is now widely accepted that GDP is not the best indicator of the health and sustainability of an economy or the wellbeing of the people that the economy supports. The carbon market creates prices that help redress this problem.

A reasonable question is how markets that trade emissions rights can help the environment, markets that are implicated in creating the global environmental problem we face today. How can a market solution correct a market problem? We must rethink our assumptions: environment and markets need not always be at odds.

GDP growth is not the source of all problems, nor is it the best index of economic success. In recognition of this, the United Nations is reconsidering its measures of economic growth and systems of national accounts. Since the turn of the 21st century, the United Nations' Millennium Goals Program has been monitoring the satisfaction of basic needs across the world — further recognition that other measures of progress are needed beyond GDP.

GDP growth is not synonymous with progress, especially where the environment is concerned. But there is an important link between poverty and environmental degradation, which can be traced to false interpretations of what constitutes a nation's comparative advantage. Because GDP and market prices more generally ignore the environmental costs of economic activity, they improperly signals what countries should specialize and trade in within the global marketplace. Countries may appear able to provide goods and services at comparatively lower costs, but this is only because market prices ignore the environmental and social costs of the production. Prices that ignore environmental and social costs mis-signal what countries should specialize in and produce for the global economy. Unrealistically low resource prices lead to overexploitation of resources and poverty. They are at the heart of the global warming crisis.[34]

The way we measure economic progress using GDP is particularly flawed for developing nations. It is inappropriate, because it is a characteristic of developing nations that have not gone through industrialization to treat their natural resources, such as forests, fisheries, fresh water supplies, and mineral deposits, as common property resources.[35] As common property resources, these countries cannot restrict access to these resources or protect them from overuse. The price for using these resources does not reflect their underlying scarcity, the damage that is caused by the processes of extracting them, or the loss of ecosystem services or values when they are removed from natural systems in unsustainable ways. What this means is that the price for these resources in developing countries is inappropriately low. This gives the false impression of a comparative advantage, and it leads poor nations to specialize in exporting resources. It has been shown that this leads to a skewed pattern of trade between the North and the South, where the South exports natural resources to the North at very low prices, which in turn leads to an overconsumption of resources across the world.[36]

Over the last 25 years, even as the United Nations (U.N.) community enthusiastically adopted basic needs as the central goal of sustainable development at the 1992 Earth Summit, the world economy increased its momentum in the opposite direction. This unfortunate trend confirmed the worst predictions about the impacts of resource trade on poverty and on the world's resources. Today, about 1.2 billion people live below the threshold of meeting basic needs. We have rapidly magnified the use of natural resources around the world, particularly in developing nations, and we have also magnified world poverty. Both results are driven by an unsustainable model of economic growth based on GDP that encourages increasing amounts of resource exports by developing nations at prices that are often below replacement costs.

Recent empirical work has shown that current measures of GDP increase as developing nations open their economies to trade, most of which is trade in natural resources. This also shows how opening up an economy to trade inevitably increases the inequality of income within the exporting nation, undermining economic growth — an unavoidable connection that was established and predicted way back in 1979.[37]

It has been shown that what matters most for growth is what the nation trades.[38] Exporting raw materials or labor-intensive products does not help countries to move out of poverty. Exporting capital-intensive or knowledge-intensive products, products that correspond to a higher level of development than the nation as a whole, is what makes export policies favorable for these nations, creating growth, wealth and overall progress. Poor countries, with their abundance of natural resources and low-skilled labor forces, specialize in the former. Rich nations specialize in the latter. At present our global trade patterns — the specializations of each country — are destined to increase the global income divide. Can the global carbon market help reverse this? The answer is yes, as is shown below.

Today we can see the results of the export-led growth policies of the last 40 years, which are based on maximizing GDP and on false comparative advantages that overrepresent the gains from trade, leading to more inequality and deprivation in exporting nations.[39] About 40 years later we face the worst environmental crisis in history and have the greatest number of poor people on the planet, both of which are caused by runaway overuse of natural resources.

We must undo all of this; we must redress the world's overuse of natural resources and the attendant runaway poverty and degradation in the developing world — the two problems are intimately connected. The whole situation appears to be a misunderstanding of gargantuan proportions, a cognitive dissonance between orthodox economic theory, history and practice. As Nicholas Stern said, climate change represents the largest externality of all time. The carbon market prices can help correct this monumental error.

Trade vs. Environment: A False Dilemma

The traditional so-called trade-off between economic development and the environment does not exist. It is illusory at best, and deeply wrong and damaging at worst — it portrays a false choice. The entire issue of trade and the environment needs rethinking, because sustainable economic growth is actually consistent with sustainable trade strategies. Appropriate policies for trade and for the environment reinforce each other. Exporting wood or petroleum is not the same as exporting software or carbon credits.

After decades of using theories of economic development that encourage growth based on exports of resources, the two-way relationship, where trade is viewed as encroaching on the environment, and environmental issues on trade, came to the fore at the trade negotiations of the World Trade Organization's (WTO) agreement. Just like the conflicts between rich and poor countries over climate change, the process exposed profound differences in perspective, and even clashes of interest, between the rich industrialized nations and the developing nations. The clashes between the positions of the North and the South continue today. What is it about the trade and environment nexus that polarizes public opinion?

It must be understood that talking about the North and the South is an oversimplification. We see this in the climate negotiations. The issue of whether to cap emissions in China is very different from the issue of whether to cap emissions in Kenya, Nicaragua, or Laos. The North and the South are far from being large blocks defined by common interests. The U.S. and EU continue to differ on fundamental issues of trade and the environment, such as agricultural subsidies, genetically modified organisms and — as is the subject of this book — the control of greenhouse gas emissions. Similarly, the South represents nations at different stages of development and different interests. Brazil is very different from Bolivia, Nigeria, or Cameroon. China is very different from almost all other developing nations.

Even so, and particularly on issues related to the environment and trade, the North–South dichotomy continues to be relevant. It helps to explain the problems in a way that will help to find a solution. Earlier in this chapter we gave a reason for this: the central issue, the core of the global environmental dilemma, is the way human societies around the world organize property rights and price natural resources. This is the source of the global environmental problems we face. The conflict between trade and the environment has the same origin.[40]

In the North, natural resources are generally held as private property and traded as such, while in the South they are held as common property. Resources — forests, water, or oil — are often called the "property of the people" in the South. This North–South difference is at the foundation of the entire environmental dilemma, as well as the source of the pattern of natural resource trade in the world economy.

The North–South pattern of trade is responsible for the worst environmental problem of our times. Global warming arises from overuse of fossil fuels, which in turn stems from extremely low fossil fuel prices in the global market. Global warming would not exist if fossil fuel prices were substantially higher, in which case we would be using other available forms of energy. But low fossil fuel prices, determined directly by international markets, are what we have been used to. Oil is a global commodity and its price is a global issue. Petroleum has been very inexpensive in recent years because it is exported from developing nations that offer their natural resources at too low a price. More realistic — substantially higher — market prices for oil could solve the problem. But markets have their own ways of determining prices. The price of oil within the global markets depends on the functioning of natural resource markets. In order for a market to reflect true costs, there must be well-defined property rights on natural resources in the exporting nations — the developing nations.

The negative connection between international trade and the global environment can be overcome through the development of global property rights on natural resources. The essence of the Kyoto Protocol is precisely to assign limits on the use of the planet's atmosphere. These are property rights. Furthermore, appropriate systems of property rights on global resources — biodiversity, the global airwaves, the planet's atmosphere, and the water masses of the world — can be created at a global scale, and used as a practical tool for leveling the playing field between rich and poor nations. They can also be used to design effective and policy-relevant solutions to resolve the conflict between the rich and the poor nations.

It is worth pointing out that the positions of the North and the South on the issue of trade and development have changed dramatically over time. Traditionally, as we saw before, the South resisted liberalizing international trade for fear of the North's domination of the global markets. Almost paradoxically, over time the North and the South have shifted places, each taking the side previously held by the other. Initially developing nations feared trade and liberalization, which could result in deforestation and poverty at the hands of powerful Northern governments, governments that represent the interests of large corporations which are unwilling to honor their commitments in trade negotiations.

Developing nations now favor international trade more than the industrial nations. In the WTO, developing nations insist on free trade for their products. Meanwhile, industrialized nations are now seen as protecting their markets, agricultural and labor, and are against outsourcing. In that sense they view each other as antagonists. Labor organizations in the rich nations have found common cause with environmental groups that are concerned with imports from developing nations. Labor groups often oppose exports from developing nations that can result in deforestation, climate change, loss of biodiversity, species loss and other forms of environmental degradation.

While the Northern and Southern interests coincide in blocking many negotiations, they disagree on who is responsible, who are the villains and who are the heroes. Some believe that international corporations put profits before people and hire cheap labor, thereby causing unemployment at home. These are, therefore, the bad guys. For the environmentalists, instead, the bad guys are the Southern governments along with the greedy multinationals. All this seems out of focus. The key issues are property rights — namely putting limits of the use of global resources — along with the atmosphere, the biosphere and the hydrosphere.

The emphasis we give to the issue of property rights is not surprising. The issue of property rights is certainly not new, but what is different and new is the emphasis on global property rights to resources, rather than the more familiar issues of national or local rights to land such as land reform.

Indeed, issues of property rights have always played a key role in economic thinking. In the 20th century they were used to separate capitalism from socialism. Capitalism was seen as an economic system based on individual private property rights over the means of production — capital — while socialism emphasized common or social property rights over capital. The two political systems, capitalism and socialism, differ precisely in their views of what is the best property rights regime for the inputs of production such as capital. Capitalism says capital should be private property, socialism says it should be owned as common property. All this is well known — the debate between capitalism and socialism is really somewhat dated and not relevant to the environment. China, a socialist nation, has severe environmental problems, and overexploitation of natural resources is associated with international markets, rather than with capitalism.

The issue of property rights remains, however, current in a radically different way. The world economy can be best viewed today as divided, not into socialist and capitalist nations, as it was in the early 20th century, but rather into the North and the South, the rich and the poor nations. The South is composed mostly of pre-industrial or agricultural economies who have not completed their industrial revolution, and the North of post-industrial economies. In both types of economies capital is not the most relevant input to production; it is not the main issue. The issue is no longer "who owns the capital?" but rather, "who owns the natural resources?" and "who owns knowledge?" Natural resources and knowledge are the critical assets of the 21st century. Both are generally global public goods. The issue relevant for the global environment is the global property rights to global natural resources: which countries own the rights to the resources that will be critically important in sustaining future welfare. This changes the underlying premises of capitalism and socialism about who owns capital. Today, both capitalist nations, such as the U.S., and socialist societies, such as China, are burdened by the same environmental dilemmas.

Pressing Concerns

A skeptical reader may ask why this problem was not detected before — why is the issue of global property rights on resources emerging only now? The reason is simple and can be best seen by analogy. We did not worry about the property rights to use roads, namely traffic light systems, until there was enough traffic. The first settlers in America did not worry about the property rights to land — it was free until it became scarce. And we never worried about global rights to natural resources, the atmosphere of the planet, its water bodies and its biodiversity until they became scarce, until human populations increased sufficiently to endanger these resources. In the entire history of our planet, human populations have never been so large: there are 7.2 billion people on the planet today and there are predicted to be 9 billion people by 2042.[41] The population growth we experience now was unprecedented before the 20th century. Because the situation is relatively new, our historical institutions are ill prepared for the change. We lack global organizations to deal with the new global challenges. Now we urgently need to organize global society with respect to its use of natural resources. This is similar to the need to organize traffic as the number of vehicles on the road increases rapidly, producing congestion. The two needs follow the same rationale: to avoid unnecessary conflict and strife, costs, suffering, and deaths. This is why global property rights on resources were not an issue until now; the world population has never been so large.

Natural resources such as petroleum, water bodies, and the atmosphere of the planet are today more important than capital in a global context. They determine the main economic issues of globalization and the attendant environmental risks, as well as the global conflict between trade and the environment. Capital is no longer the main input of production in advanced societies as it was in the industrial societies at the beginning of the 20th century. Nor is capital the main determinant of production and trade in developing nations. In preindustrial economies, which we call the South, the property rights that matter are predominantly those used for owning and trading land and agricultural products and, more generally, those for owning and trading natural resources. Similarly, in the post-industrial economies that buy resources from the South, the main input of production is now knowledge rather than capital. For these reasons, the explanations given here for environmental and trade issues, and the solutions proposed for these issues, focus on a different type of property rights than those that were important at the beginning of the 20th century. Property rights on resources and knowledge are more relevant in today's world, one that is divided between the North and the South.

In summary, the debate about the environment today is not over socialism or capitalism but rather between two other forms of economic organization — agricultural and post-industrial societies that are connected through international markets. The environmental dilemma cuts through and across conventional political divisions of left and right, capitalism and socialism. This has been confusing to many who persist in holding onto somewhat outdated left–right forms of thinking. Conserving the environment is important for both sides. The basic environmental issues we face are due to the fact that natural resources are exported and overextracted in the South and are imported and over-consumed in the North. This is the relevant dichotomy that we must address if we want to understand and resolve the global environmental dilemmas of our times: global warming, ozone depletion, a lack of drinkable water, and destruction of the complex web of species that make life on earth. We must deal with the economic foundations of a market-based relationship between the North and the South.

The leading environmental issue is the over-extraction and the overconsumption of natural resources across the world. The over-extraction of natural resources in the South leads to the overconsumption of these resources in the North. At the end of the day, this is what the global environmental problem is all about. Think of it this way: if the North had significantly reduced imports of petroleum and the South had significantly decreased oil production and extraction of forest products, thereby allowing the number of forests in the world to significantly increase, then the global warming problem would not exist. It would disappear. And many other global environmental problems would be resolved or greatly improved. The majority of the world's biodiversity threatened with extinction today involves species that live mostly in the world's forests and their surrounding areas and water bodies; they could be sustained if their ecosystems remained intact.

The process of using global limits on property rights to resolve the issues of trade and environment has begun. Its beginnings were humble and largely misunderstood, but the process is so important that it begs for clarification of the policy tools that can accelerate its adoption and use.

Global property rights on resources are the main ingredient and the most distinguishing feature of the Kyoto Protocol. The Protocol created a global system of property rights on the atmosphere, a natural resource that is necessary for survival in our era of globalization.

International agreements such as the Kyoto Protocol, with its groundbreaking system of global property rights on the use of the atmosphere, hold the key to the future. They can resolve and harmonize the worst conflicts we face in the areas of trade and the environment. The Kyoto Protocol represents only a beginning, a template for what is to come. However, if one could design the global economy today — to ensure a better future for billions of people across the planet — one could not do much better than to use this template as a blueprint of what is needed, of things to come.

The Kyoto Protocol is more than a system of allocations of emissions limits, or property rights, on the use of the atmosphere — who can emit what, and the trading of these rights among nations. It also contains an asymmetric treatment for the North and the South, for the poor and the rich nations. By purposeful design, this is a global market that favors poor countries. They own most of the environmental resources in the world and they can use these to sell CDM credits to industrial nations. This is a step in the right direction, both for the environment and for eliminating global inequality in a way that limits carbon emissions and helps resolve climate change. This is why the Kyoto Protocol's carbon market can pass the litmus test of even the most ardent market and trade skeptics. It deserves to be given a chance.

6

Implementing the Carbon Market and its CDM

The Kyoto Protocol is far from perfect. It nonetheless achieved something that is important for us all, which most people had thought was impossible: an agreement to cap global greenhouse gas emissions from the industrial nations, the largest emittors at the time. It is the first time in our history that we have agreed to limit our carbon emissions and change our energy use, which is a way to avert climate change.

Even if we wanted to reinvent the wheel and try to negotiate yet another international agreement to reduce emissions from scratch, we would still be drawn to follow the Kyoto route. Why? Because Kyoto has extraordinary features that are needed, which are often misunderstood despite their benefits for humankind.

What are these magical features? One basic feature is to cap global emissions. A second is the carbon market, which can resolve global warming at no net cost to the economy. This may seem unbelievable, but it is true. This feature has not been observed until now and needs to be highlighted. Many concerns have been expressed about the costs of averting global warming, which are estimated to be about 1% of the world's gross economic output. This is the basis for opposition to efforts to combat global warming. But these concerns are put to rest by the carbon market in one fell swoop.[1] The carbon market's trading operation requires not a single dollar donation, nor taxes, nor subsidies — nothing at all. In fact, it is the other way around: the carbon market generates revenue and can stimulate the world economy.

Finally, the carbon market can create abundance and provide jobs in developing countries, restoring ecological function on a massive scale. Like the Green Belt, it would benefit rich and poor.[2]

These are bold statements that will not go unchallenged. How does it all work?

Kyoto, Global Wealth Creation, and Sustainable Development

Kyoto works by the magic of the market. Yes, that is all there is. But do not underestimate what it means and what has been achieved. The market is a powerful institution that creates wealth and represents democratic objectives like nothing else. But the market has been badly undermined so far. The 21st century inherited international markets that were oblivious to our social and physical limits. This is fair enough because, until recently, human societies were relatively small, and we could use as much of the planet's atmosphere as we wanted at no cost. But we have reached our limits. We have collided with natural constraints on our use of the atmosphere.

In creating emissions caps, the Kyoto Protocol changed all that. Fast. It gave the market the right signal: scarcity is real. Our scarce resources are rare and valuable. Services and goods are now known to be less valuable than working ecosystems. We cannot continue to use the atmosphere as a dump — there are limits to that. Many indigenous cultures share this view. Why are so many people in industrial nations unaware of this reality?

After the Kyoto caps on emissions were created, market prices emerged from trading these caps. The carbon market puts a price on emissions; today, this is low because of a lack of emissions limits after 2020, about $12 per ton using the European market. This means that the cost to society of emitting one ton of carbon is roughly $12.[3] After that, the world markets cannot be the same anymore. No longer will we treat the atmosphere as an unlimited resource, a depository for all of the waste that our power plants, vehicles, and factories produce. It looks simple and it is simple. It is just a matter of dollars and cents, you might think, and what is so great about that?

Not so fast. This is more than dollars and cents. In creating the carbon market, country-by-country, we have accepted limits on our use of the global atmospheric commons. This is a tremendous feat, for it marks a turning point in our relationship to nature. It symbolizes our emerging new consciousness about nature's limits and its vulnerability to human abuse. Perhaps, more importantly, it signals the start of a new era of global cooperation. If nations can overcome their differences to accept their shared responsibility for the health of the planet and the wellbeing of future generations, we may be able to find solutions to the other serious crises confronting the global community. In this way, although climate change presents an enormous challenge, by creating the carbon market we have transformed failure into an opportunity.

The Kyoto Protocol forged cooperation between rich and poor nations, business and environmental interests, in ways that no other international agreement or market relationship has been able to do thus far. But it is just the beginning. The full potential of the carbon market has yet to be unleashed. On all fronts, time is running out. We are running out of time to save ourselves and we are running out of time to prevent catastrophic climate change.[4]

The question is no longer whether or not to combat climate change. In 2015, 195 nations agreed that there is no greater threat to civilization than climate change. It is priority number one. This was the 2015 Paris Agreement. But it had no mandatory features, no teeth. The question now is how the Kyoto framework can be continued after 2020 and made effective.

If Not Now, When

We are all in this together, and we are aware of the urgency. Yet, the first reaction one hears is that each country should try to resolve its own environmental problems and avoid international complications. Why not? Why could the U.S. not create its own carbon market, why could it not limit its own emissions? Why do we need international cooperation and cumbersome international agreement? Why not go it alone? Why, indeed?

If a country adopts a go-it-alone approach to emissions reduction, total emissions will not fall sufficiently to avoid the risks of climate change. It is as simple as that. Think of it this way. The U.S. could refrain tomorrow morning from emitting a single ton of carbon. Yet, emitting nothing itself, the U.S. will still be the victim of climate change and will still suffer the consequences of global warming due to the actions of others. This is because CO_2 concentration in the atmosphere is the same all over the world, whether in Madrid or New York. This is a physical property of CO_2. In fact, as we saw in Chapter 1, the Organization for Economic Co-operation and Development (OECD) estimated in 2008 that a single U.S. city such as Miami, Florida, stands to lose $3.5 trillion in property damages due to climate change — the horrific hurricanes of 2017 underscore this fact. Just one city will suffer trillions in losses due to the actions of other nations. Why does this all happen?[5] Because the U.S. emits only about 20% to 25% of global emissions. That is all. The other 80% plus comes from outside U.S. territory and is enough, more than enough, to precipitate climate change. Carbon emissions distribute uniformly over the entire planet. It is the ultimate equalizer. Without immediate cooperation we are all doomed together.

Yet, if countries cooperate as part of an international treaty, they still need to agree on how to share the burdens of climate mitigation between them. This is exactly what Kyoto stems from: an agreement on limiting greenhouse gas emissions country-by-country. Let us repeat what the Kyoto Protocol has done: mark agreement on and assign emissions limits to the various industrial nations in the world. Now, clean carbon negative technology and ecological restoration programs can be added to the agenda.

Once emissions limits, industrial cleaning and ecological restoration have been agreed upon, nation-by-nation, the carbon market is a natural and desirable way forward. It allows each nation the flexibility to be above its goals one year and below it another year, remaining within its limits on the whole. Once you accept that flexibility, you are back to the Kyoto Protocol structure in its full regalia. Kyoto allows each nation to have its own internal control system — carbon taxes, markets, etc. There is no way to escape this simple truth.

Why Honor the Kyoto Agreements?

Kyoto has already been negotiated and it took years to do so. We need to decrease emissions limits further, improve the Clean Development Mechanism (CDM), plant billions of trees and suck up CO_2 by the gigatons. But what other mechanism has the potential to resolve competing interests as effectively as Kyoto through the carbon market? Through it, the international community can honor the commitment of fairness to developing nations that it made in 1992 at the Earth Summit. It is possible for the international community to have its cake and eat it too — fairness and climate change prevention.

We know that the Kyoto carbon market can help unite the business interests in rich nations with the clean development needs of poor nations. It has done so already through the CDM, which works alongside the carbon market. The business community defends the CDM against the European parliament, as do the environmentalists. But to reiterate, the Kyoto Protocol is a market-orientated agreement with three additional properties that make it truly unique:

(1) It creates new wealth to pay for emissions reduction, as clean nations can sell rights to emit. It can resolve global warming at almost no cost to the economy, thus contradicting the dire warnings that we have heard from many of the world's experts about the costs of averting global warming.
(2) Through the CDM, it can foster desirable changes, clean technologies, and sustainable economic development across the world. It can support and help fund developing nations' ecological restoration.

(3) It can help alleviate poverty in developing nations by transferring wealth from rich to poor nations for clean productive projects, to the benefit of all countries.

The Kyoto Protocol achieves all this using a free market approach, without taxes or regulations, except for the firm limits it places on carbon emissions. This is performance beyond anything we have seen so far. These claims deserve to be carefully argued and explained.

Wealth creation

The magic of the carbon market is that it pays for itself. Fiscal conservatives should welcome the news that the carbon market is a fully self-funded institution. Other than the cost of participating in global climate talks, which establish the rules for international emissions trading, it requires nothing additional from countries' governments.

As we have already discussed, the important thing the global carbon market does is attach a price to carbon emissions. The current price of carbon in the global market is roughly $12 per ton and could rise to $20 per ton in the 2020s as new mandatory emissions limits come into effect. The world economy currently produces the equivalent of over 32 gigatons of carbon dioxide per year. If we charge emitters $20 for each ton of carbon they produce, the carbon market can generate $640 billion a year. This is equivalent to 0.8% of global gross domestic product (GDP). This is an enormous new source of wealth that the carbon market creates. We can tap into this wealth to pay for global emissions reduction. In reality, the market will do this all by itself. Whoever can reduce carbon will sell their carbon credit in the carbon market, and will be paid for their work.

In truth, the wealth has always existed. It was the planet's wealth — the atmosphere and its capacity to regulate global climate. Before the Kyoto Protocol, emitters effectively claimed the planet's wealth — our common wealth — for their own private use. By pumping carbon and other greenhouse gases into the atmosphere, they drew down our wealth stock. But now, thanks to the carbon market, we have the ability to take back this wealth and reclaim it as our common asset. It is important that we use the global carbon market to distribute this wealth across countries. And we must do it in a way that favors poor nations, favoring us all.

For many years we have been wrapped up in questions of what it will cost to save the world from climate change and whether we can we afford it, and we overlooked the potential for emissions reduction to pay for itself. As we saw in Chapter 2, the Intergovernmental Panel on Climate Change (IPCC) surveyed the expert literature on climate economics and found that the estimated costs of preventing climate change range from 1–4% of global GDP each year. The influential Stern Report to the U.K. government, by former World Bank Chief Economist Sir Nicholas Stern, estimated that the world would have to spend 1% of global GDP each year to protect against damages, as great as 5–20% of global GDP each year, in the future. Fortunately, the global carbon market can generate about $900 billion each year — sufficient revenue to offset the costs of protecting the future against global climate change. The numbers make sense. World GDP is about $77 trillion today.[6] It could cost 2% of global GDP, roughly $1.54 trillion each year, to prevent catastrophic climate change. And the carbon market can create a new source of global wealth, $900 billion, which is equivalent to 0.8% of global GDP each year, and which will be paid to those who reduce emissions. The net effect on the global economy is zero. Some people will be worse off, and some will be better off. The bad guys, the over-emitters, will pay, and the under-emitters will receive their money. But the net effect on the global economy is zero.

Carbon prices should go up in the future, as international trading increases and countries adopt more aggressive timelines for reducing emissions. By some estimates, if emissions limits are extended post 2020, the carbon market of the Kyoto Protocol will reach trillions of dollars over the medium term, but for as long as the carbon market operates it will facilitate the level of emissions reduction that is required to avert catastrophic climate change.[7] This presupposes, of course that technologies are there to reduce carbon emissions — the same assumption made in the Stern Report. We have shown how carbon capture and storage, including new negative carbon technologies, can be used for this purpose. The price of carbon may adjust to reflect the technological reality of what it costs to reduce a ton of carbon. If the cost for reducing carbon emissions decreases, the market price will go down. This is the magic of the market.

So what does all this mean? It means that using the carbon market to penalize emitters and require them to pay for emissions will generate enough income to offset the projected costs of preventing climate change. Compared to the commercial reinsurance costs for catastrophic risks discussed in Chapter 2, 2% of global GDP to insure the future from catastrophic climate change seems fair and prudent. The carbon market gives us a way to pay for it. The carbon market guarantees that the emitters will foot the bill. They will pay the premium for all of us. And the market ensures the efficient allocation of those dollars to achieve the task.

Still not convinced? Think of it this way: at the global level, the benefits of preserving our climate system outweigh the costs. According to the Stern Report, the costs of global warming are equivalent to losing 5–20% of global GDP now and forever. It is true that we may not be able to appropriately and accurately measure those benefits. Those of us who believe in the science of climate change understand the benefits of preventing climate change. The lives saved and property damage avoided, are well worth the investment. According to the IPCC Fourth Assessment Report, this is about 1–4% of global GDP. Preventing climate change then generates a global benefit that equals or exceeds the associated global cost. It increases welfare by achieving a more balanced combination between fossil fuels uses and environmental quality. This is the aim of the international community, at no extra cost to the global economy.

One must be careful to understand this claim. To say that a successful international effort to avoid climate change can generate a net welfare gain, a global benefit in excess of global cost, is not the same thing as saying that everyone everywhere will be better off because of the carbon market. The carbon market, as we saw, will generate winners and losers: all markets do. It may hurt some businesses that make large profits generating carbon emissions, such as coal power plants and some economies that prosper from the export of fossil fuels. Now, new profit opportunities and new development prospects can be created where none previously existed. This is as it should be. Create the right incentives for using and cleaning up the planet's atmosphere. No surprises here. But as long as the gains to some sectors and some people offset the losses of others, there will be no net loss to the global economy from our efforts to combat climate change. That is the magic of the carbon market.

In the U.S., the political and academic debate over a cap-and-trade program to reduce U.S. emissions is well underway. The issue is whether to use a cap-and-trade mechanism or taxes, and how best to distribute the revenues that the initial sale of carbon permits will generate. The economics literature has shown that it is possible to shield the most economically vulnerable households in the U.S. from the costs of emissions reductions. Some noteworthy work demonstrates that, if the U.S. distributes the revenues from carbon sales on an equal per capita basis, about 60% of U.S. households would benefit in monetary terms.[8] The "dividend" payment that each American would receive would represent her/his claim on a common asset — the U.S. share of the global atmosphere. A carbon market program in the U.S. has the capacity to generate enough wealth to raise the incomes of almost two-thirds of Americans. This is truly extraordinary.

As we will see, the potential for redistributing wealth exists on a global scale as well. The distribution of emissions rights and the CDM are how the world can distribute the benefits of the global carbon market. This is how the distribution shall benefit those nations that emit less while penalizing those that emit more.

Let us be clear. Even though preventing climate change can make every country better off, there is no guarantee that every country will be better off. There are no guarantees in nature. A guarantee is a human invention. The global carbon market creates the right potential and incentives, but it is still up to us to decide how we distribute the benefits of the carbon market. Imagine that the carbon market is a successful recipe for baking a larger pie for the global community to share. A larger pie means that it is possible for all of us to have a bigger slice. There are many ways for us to divide the larger pie: we can divide the pie in a way that is equitable, giving those countries that have only consumed small slices in the past much larger slices in the present. Or we can divide the pie such that the rich countries' slices get bigger, while other countries' slices shrink or stay the same. Now that we understand why the carbon market is the recipe for the future, we can agree to divide the pie equitably between us, so that we all benefit. We use the carbon market to promote sustainable development and to close the global divide.

Fostering sustainable development and overcoming poverty may seem too much to ask for but it is possible. In our view it will happen. This is the magic possibility of the Kyoto carbon market when we have responsible and committed stewards at the helm. China is implementing the carbon market and so are several U.S. states including California.

Fostering sustainable development

Trading between developed and developing countries can be tricky. Selling one's rights to emit carbon dioxide today means selling one's ability to burn coal and other fossil fuels. Most industrial economies would screech to a halt if they were unable to burn fossil fuels today, unless alternative energy sources and technologies were available. By selling their rights to emit, developing countries could be selling their rights to industrialize.

The global carbon market can help solve this dilemma. The richest countries have money, but an environmental deficit: they emit about 50% of global greenhouse gases even though they make up less than 18% of the world's population, according to the OECD. The developing countries are in the opposite situation: they have a credit in the environmental account, but a monetary deficit. They emit less and they house most of the world's remaining forests and biodiversity. This is where the real trading value lies, not in goods and services. There are natural gains from trade between the two groups of countries.

How can the carbon market provide incentives to protect, rather than to destroy, the world's forests? The carbon market gives countries a way to capitalize on the value of their environmental resources. These are the most valuable resources known to humankind because they are needed for our survival, air, water and food. Yet, at present, multinational corporations only realize their value by destroying them: a forest that preserves biodiversity and contributes to the atmosphere's quality is destroyed to sell the wood of its trees for plank or pulp, or burned to give way to arable land. Carbon markets can help realize the value of environmental assets without destroying them. They can balance out the position of large and small traders by offering a neutral trading base for all. They can provide an anonymous process where several small sellers can meet a few large buyers. Carbon markets can be an important part of restructuring the global financial infrastructure to meet the needs of today's mature industrial economies, as well as those of newly industrializing countries.

Even though the developing nations have no emissions caps in the Kyoto Protocol, and therefore cannot trade directly in the carbon market, they can still participate and benefit from the carbon market. This is because of the Clean Development Mechanism (CDM), which encourages investment in clean technologies in developing nations, allowing developing nations to benefit indirectly from carbon trading. The CDM is the crucial link between emissions reduction and the broader goal of sustainable development. The Kyoto Protocol stipulates that CDM projects must be able to demonstrate tangible sustainable development benefits for the host country. A CDM project belonging to France could be located in Nigeria, just as embassies of one nation are located in another.

We mentioned the CDM before but now we will explain how it actually works. When an industrial nation, individual, or institution invests in a project that reduces emissions inside a developing nation the investor is given a credit based on the amount of carbon that is actually reduced by the project. This carbon credit can then be traded in the carbon market. For example, a project that is proven to reduce carbon emission by one million tons will be awarded carbon credits that can be traded in the carbon market at the current carbon price of $12 per ton. This increases the project's profitability significantly, by decreasing costs by $12 million. This way the CDM produces strong incentives for the development of clean technologies in developing nations and encourages investors in industrial nations to finance such projects. It ensures a flow of investment dollars from North to South, while transferring the technologies of the post-carbon economy to developing nations.

The CDM changes the profit equation in favor of using clean technologies in developing nations. To see this, consider for example two projects in developing countries that are identical in every possible way, except for the technology they use. One project emits 10 million tons of carbon, the other project emits no carbon at all. Before the Kyoto Protocol attached a price to carbon emissions, investors had no incentive to choose one technology over the other. But with the CDM, the project using the clean technology, the one that produces no carbon emissions, becomes $12 million (using current carbon prices) more profitable than the other project.

The CDM is an important and potentially transformative market incentive embedded in the carbon market of the Kyoto Protocol. During 2006, the first year in which the carbon market traded, about $8 billion in such projects were carried out in developing nations. The Kyoto Protocol carbon market, which is the EU Emissions Trading Scheme, has already led to transfers of approximately $215.4 billion to poor nations for CDM projects. These projects account for about 20% of the EU's required emissions reduction.[9] As the Kyoto limits on emissions come to an end, the CDM weakens. But mandatory limits can and should continue.

One need only explore the scale and scope of implemented project activity under the CDM to understand its

Table 1 Expected CERs

	Expected CERs until End of 2020*
7,655 are registered	7,784,211,709
11 are requesting registration	5,593,684
194 are pending publication	238,076,024

Note: * Assumption: no renewal of crediting periods. CER = Certified Emissions Reductions.
Source: United Nations Framework Convention on Climate Change (UNFCCC).

transformative potential for the global economy. There are already well over 7,500 registered CDM projects worldwide, just under 50% of which are produced in China alone (see Table 1 "Expected CERs"). Each Certified Emission Reduction (CER) is worth one ton of carbon emission equivalent. As of 2008, there were over 7,000 CDM projects in the pipeline, at various stages of approval, with the potential to produce 2.7 billion CERs between 2006 and the end of 2012.[10] More than 70% of the projects are aimed at developing renewable energy resources in the host country (see Figure 1 "Distribution of CDM Registered Project Activities by Scope"). The data reflects the end of the mandatory emissions limits, which require extension.

Each CDM project must qualify through a rigorous public registration and issuance process, the purpose of which is to ensure "real, measurable, and verifiable emissions reductions that are additional to what would have occurred without the project."[11] The CDM Executive Board, which answers to the nations that have ratified the Kyoto Protocol, oversees the verification and registration of CDM projects.

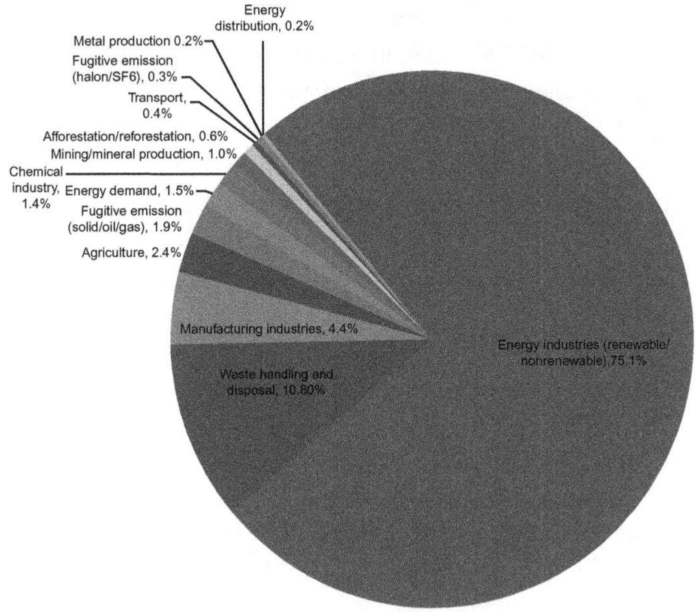

Figure 1 Distribution of CDM Registered Project Activities by Scope
Source: Using data from https://cdm.unfccc.int/Statistics/Public/files/201506/proj_reg_byScope.pdf.

Clean development projects worldwide[12]

To better understand the real benefits of CDM projects, we describe a handful of important case studies below. These are the official descriptions of the projects and their sustainable development benefits, as reported to the CDM Executive Board for project approval and certification. (See Chapter 6 for details on how financial assistance is given and how credits are awarded.)

Inner Mongolia wind farm project

Host Country: China
Associated Country: United Kingdom

The objective of the Inner Mongolia Wind Farm Project is to generate electricity from renewable wind resources. The project will install 33 wind turbines. Total installed capacity will be 49.5 MWh. Once the project is put into operation, the annual average power delivered to the grid is expected to achieve a level of 115,366 MWh and the expected annual emissions reduction is 120,119 tons of carbon emission equivalents. Clean and renewable energy will be utilized by the project for power generation, which is environmentally and socially beneficial. The project will contribute to sustainable development in the following ways:

- Reduction of greenhouse gas emissions and other pollutants.
- Promotion of domestic manufacture and development of a wind power industry in China.
- Increased local revenue and employment opportunities.

Fuel switching project of the Aqaba thermal power

Station Host Country: Jordan
Associated Country: United Kingdom

The project is to switch from the use of oil to the use of natural gas (NG) at the Aqaba Thermal Power Station (ATPS) in Aqaba, Jordan. ATPS is the largest power plant in Jordan. The fuel switch is from Heavy Fuel Oil (HFO) to NG, and the capacity of the plant is unchanged as a result of the fuel switch. The project is estimated to reduce an average annual amount of 397,163 tons of carbon equivalent emissions over a 10-year crediting period. ATPS initiated this fuel switch because of the plant's negative environmental impacts, which are mostly gaseous and a result of HFO combustion for electrical power generation, and because of Jordan's ratification of the Kyoto Protocol and potential CDM benefits, which were considered from the beginning of the project to make it financially viable (despite unfavorable relative fuel prices). The fuel switch will benefit the environment, and contribute to sustainable development as follows:

- Support of the local economy, which is dominated by tourism and, therefore, benefits greatly from reduced pollution.
- Reduction of shipping/trucking of HFO, with reduction of related traffic and pollution (NG will be imported from Egypt via a submarine pipeline in the Gulf of Aqaba).
- Reduction of greenhouse gas emissions and diversification of Jordan's electricity production with a leaning towards "cleaner" power.

Biogas capture and regeneration at Lekir Palm Oil Mill

Host country: Malaysia
Associated Country: Sweden

(Palm oil production is at the root of rainforest destruction. This example has been given only to illustrate the CDM possibilities.)

The proposed project was to be implemented at Lekir Palm Oil Mill in Malaysia. The wastewater from the mill is treated through a ponding system. These conditions result in anaerobic conditions within the ponds and biodegradation of the organic content in the wastewater leads to the generation of methane. The proposed project activity is to cover two of the existing open anaerobic open ponds to capture the methane-rich biogas. The treated effluent will then be channeled into the existing subsequent ponds for further polishing. The captured biogas will be combusted in dual fuel generators (fired with a mixture of diesel and biogas) to generate electricity for the project activity while the excess biogas will be flared on site. This process will benefit the environment, and contribute to sustainable development as follows:

- Reduction of the greenhouse gas emissions from the open anaerobic ponds.
- Improvement of air quality through the reduction of odor.
- Creation of job opportunities for the local community during project implementation.
- Enabling the transfer of knowledge and technology on biogas generation, treatment and the utilization of biogas in generators, which should stimulate the development of both the market for generators and the use of biogas as a real alternative energy for the local industries in Malaysia.

Yuexi Dayan Small Hydropower

Project Host Country: China
Associated Country: Sweden

The project is a water-diversion-type run-of-river hydropower project located on the upper reach of the Yangbijiang River in Yunnan Province, China. Electricity generated by the proposed project will displace part of the electricity generated by South China Power Grid, which is dominated by coal-fired power plants, and thus greenhouse gas emission reductions will be achieved. The average annual emissions reductions of the proposed project are estimated to be 97,403 tons of carbon emissions equivalents. The proposed project will promote local sustainable development by making use of renewable hydropower. Major contributions of the proposed project are as follows:

- Building the major power plant in Yangbi Yi Autonomous County will play an important role in developing local resources such as silicon, antimony, and molybdenum, which can help local poverty reduction.

- Alleviation of the shortage of electricity and improvement of the power quality of the grid to ensure smooth industrial and agricultural production and meet the need for electricity in the daily life of the ethnic minorities in the region.
- Reduction of the emission of pollutants and greenhouse gases that might otherwise be caused by coal-fired generators so as to improve local environment.
- Creation of local jobs.
- After the operation of the proposed project, the option for local people to substitute firewood with electricity, which will reduce the damage to the local vegetation and protect the local ecology.

Facilitating reforestation for Guangxi watershed management in Pearl River Basin

Host Country: China
Associated Countries: Italy, Spain

The proposed project aims to explore and demonstrate the technical and methodological approaches related to credible carbon sequestration (CCS) and pilot the viability of enhancing the livelihoods of people and the natural environment by facilitating reforestation activities in watershed areas along the Pearl River Basin. The project will generate income for the poor farmers/communities by enabling the carbon sequestered by plantations to act as a virtual cash crop for local project beneficiaries, who will gain direct benefits from harvesting the plantation as well as from the sale of carbon credits, which will in turn reduce the threats to natural forests. In addition, forest restoration in this area plays a vital role in biodiversity conservation, soil and water conservation, and poverty alleviation, while sequestering carbon dioxide from the atmosphere. The sustainable development potential of the project is as follows:

- ➢ Sequestering of carbon through forest restoration in small watershed areas and testing and piloting how reforestation activities generate high-quality emissions reductions in greenhouse gases that can be measured, monitored, and verified.
- ➢ Enhancement of biodiversity conservation by increasing the connectivity of forests adjacent to nature reserves.
- ➢ Improvement of soil and water erosion control.
- ➢ Generation of income for local communities.

Agricultural waste management system emissions mitigation

Host Country: Mexico
Associated Countries: Switzerland, United Kingdom

(Although the authors of this book do not at all approve of industrial livestock breeding, this case is mentioned only to illustrate CDM financing.)

This project proposes to apply to confined animal (swine) feeding operations in central Mexico, a greenhouse gas mitigation methodology that is applicable to intensive livestock operations. The proposed project activities will mitigate greenhouse gas emissions in an economically sustainable manner, and will result in other environmental benefits, such as improved water quality and reduced odor. In simple terms, the project proposes to move from a high-emissions animal waste management system — an open air lagoon — to a lower-emissions animal waste management practice — an ambient temperature anaerobic digester with capture and combustion of resulting biogas. The purpose of this project is to mitigate animal effluent-related greenhouse gas emissions by improving animal waste management systems. It has the potential to reduce approximately 4.3 million tons of carbon emission equivalents each year. Its contribution to sustainable development is as follows:

- Protection of human health and the environment through proper handling of large quantities of animal waste.
- Methane recovery project activity to upgrade livestock operations infrastructure and enable the use of renewable energy sources.
- Improvement of air quality (reduction of the emission of Volatile Organic Compounds and odor).
- Incentive for future farm projects that will have an additional positive impact on greenhouse gas emissions, with an attendant potential for reducing groundwater contamination problems.
- Increased local employment of skilled laborers for the fabrication, installation, operation, and maintenance of the specialized equipment.
- Establishment of a model for world-class, scalable animal waste management practices, which can be duplicated on other livestock farms throughout Mexico, dramatically reducing livestock related greenhouse gas emissions and providing the potential for a new source of revenue and green power.

Kuyasa low-cost urban housing energy upgrade, Cape Town

Host Country: South Africa
Associated Country: None

This project is aimed as an intervention in an existing low-income housing development with households in Kuyasa, as well as in future housing developments in this area. The project aims to improve the thermal performance of the existing and future housing units, improving lighting and water heating efficiency. This will result in reduced current and future electricity consumption per household and significant avoided carbon emissions per unit. Other benefits of the project activity include a reduction in local air pollution with subsequent decreases in pulmonary pneumonia, carbon monoxide poisoning, and other respiratory illnesses. A decrease in accidents and damage to property as a result of fire is also anticipated. The project activity relates to the following three interventions per household unit: (i) insulated ceilings; (ii) solar water heater installation; and (iii) energy efficient lighting. It contributes to sustainable development as follows:

- Improved end-use energy efficiency combined with the use of solar energy for water heating, resulting in measurable avoided pollutant emissions and energy consumption savings. This contributes to "energy poverty" alleviation.
- Increased use of renewable energy and improved thermal performance, such that cleaner energy services are provided with respect to local pollutants, and are cheaper than in the baseline situation. The improvements in thermal performance will moderate indoor air temperature with associated comfort and health benefits.

Substituting biofuels and municipal solid waste in cement manufacturing, Tamil Nadu

Host Country: India
Associated Country: Germany

The project activity is the partial replacement of fossil fuels with alternate fuels (de-oiled rice bran, municipal solid waste, tyre) in cement manufacturing at Grasim Industries

Limited, Cement Division South, Tamil Nadu, India. The purpose is to reduce carbon emissions in cement production by using alternative fuels. Conventionally fossil fuels, namely coal, lignite, and pet coke are used in the kiln system for clinker formation. The project involves partial replacement of the fossil fuels with alternative fuels, such as agricultural by-products, tyres, and municipal solid waste (in the form of refuse-derived fuel), all of which are lower greenhouse gas-emitting fuels. This will result in significant saving on non-renewable fossil fuel as the project has the potential to mitigate 51,932 metric tons of carbon emissions equivalent per annum. Utilization of these alternative fuels would require retrofitting of the existing facility and installation of fuel processing equipment. This type of project is not common because it is not normally financially viable without CDM financing. Its sustainable development potential is as follows:

- Substitution of lower-emission fuels for fossil fuels.
- More effective disposal of waste by using it as an alternative fuel.
- Generation of employment opportunities in agricultural by-product supply chain.
- Generation of employment for skilled and unskilled workers of the rural region.
- Creation of an additional source of revenue for farmers' agricultural by-products, which were earlier burnt in open grounds and fetched no value.

Closing the Global Divide

The climate crisis is a classic example of over-exploitation of a common property resource, often called "the tragedy of the commons." It is the global tragedy of the commons. The atmosphere is a resource that all people on the planet share. Throughout history we have polluted this resource without consideration for how it diminishes what is available to the billions of other people who we share it with. We never had an incentive to economize on our use of the atmosphere, because there was no cost for using it. The atmosphere belonged to all of us and to none of us at the same time.

With this simplistic, yet accurate, depiction of the problem, the logic of the Kyoto Protocol becomes clearer. Since no country was empowered to restrict access to the atmosphere, and no country had an incentive to limit its own use, the only chance we had of preserving the atmosphere for ourselves and for future generations was to collectively agree to a cap on global emissions. By capping global emissions, the Kyoto Protocol transformed something that we had historically treated as unlimited in supply — the atmosphere's capacity to absorb greenhouse gas emissions — into something that was finite. This acknowledged a scarcity that was previously overlooked. That scarcity now imposes a cost on our use of the atmospheric commons. For the first time in history, we pay a price for polluting the atmosphere.

What the Kyoto Protocol essentially did was create property rights — user rights — to a globally shared resource. We can think of these user rights as emissions rights; the right to emit a fixed quantity of carbon dioxide into the atmosphere. But limiting access to the atmosphere was only step one. The more difficult task for Kyoto was to allocate those user rights to nations. It did this by assigning emissions limits to industrial countries. The limits specified the percentage reduction in emissions that each country had to achieve by the end of the first commitment period. Implicit in each country's emissions limit was its share of global emissions rights. The lower a country's emissions limit, the fewer user rights it was granted, and the more it was required to cut its emissions.

To fully understand the conflicts threatening the future of the Kyoto Protocol, it is important to recognize that the allocation of emissions rights is a form of wealth transfer between nations. The absence of emissions caps on developing countries' emissions essentially means that they have unlimited access to the atmospheric commons. In this sense, the Kyoto Protocol capped only what was then the largest portion of global emissions, the portion coming from industrialized countries. By granting developing countries unlimited user rights to the atmosphere, the Kyoto Protocol allowed, through the CDM, a huge incentive for clean productive investments in poor nations, one that is virtually unrivaled in international affairs. Emissions rights are highly valuable commodities, especially when those rights can be traded in a global carbon market. The allocation of emissions rights is a potent tool for increasing clean productive investment between nations and we can use it in a deliberate way to close the global income divide and prevent global warming.

There is a somewhat surprising feature of the carbon market that makes it favor equity. Unlike all other markets, the carbon market has to protect poor countries as a condition of market efficiency. To achieve efficacy in the carbon market requires a measure of fairness. There is no other market for which this is true. In all other markets, outcomes can be efficient but highly unfair at the same time; this is the reason why many people mistrust markets. That cynicism need not carry over to the carbon market.

The "commodity" that carbon markets trade is, in reality, different from any other commodity ever traded: it is a global public good. As a result, there is an important link between each nation's right to emit and the efficiency of the market. This has important implications for market behavior.[13] Because carbon dioxide is distributed uniformly and stably around the planet, the concentration of carbon dioxide is the same for everyone around the world. This concentration cannot be chosen independently by each trader — we all face the same concentration of carbon dioxide. This uniformity is one way of characterizing a global public good. Classic public goods are those whose supply is the same for everyone involved, for example the armed forces, bridges, a national constitution, and school systems. Markets that trade the rights to use public goods are different from standard markets for private goods, such as corn, machines, houses, stocks, and bonds. In standard markets, traders decide how much of different goods to consume, and they do so independently from each other. Each trader identifies their optimal level of consumption of a good based on their preferences and what they can afford. The optimal level of consumption for one trader is not necessarily the same as the optimal level of consumption for another.

CO_2 distributes uniformly over the whole world. Therefore the Kyoto Protocol specifies the same carbon dioxide concentration in the atmosphere for all countries. Given the stark differences in countries' incomes and what they can afford, the concentration that the Kyoto Protocol specifies is not necessarily optimal for all countries. Indeed, it is most likely beyond what most developing nations can afford. Developing countries use little energy and produce few emissions compared with the rest of the world. They confront more immediate short-term demands to satisfy basic needs, such as those to feed, clothe, and educate their populations. For these reasons, it is inefficient to require developing countries to purchase more global emissions abatement than they can afford at this point in time. But this is exactly what the carbon market of the Kyoto Protocol avoids — which it has to do since emissions abatement is a global public good and we all share the same atmosphere.

The only way to solve the problem, to make it efficient or optimal for developing countries to afford Kyoto's level of global emissions reduction, is to increase clean investment in developing countries and allow them to reduce emissions to reach the levels agreed to for the world as a whole. We could do this directly, through lump sum payments to developing countries along the lines of what international aid agencies try to do. We could also issue "side payments" to induce developing countries to accept emissions caps.[14] But the magnitude of the wealth transfers that would be required in these cases is far greater than any income transfer we have seen before. One can imagine the uproar in rich nations if governments began transferring huge sums of wealth directly to nations such as India or China.

The magic of the carbon market is that it avoids these issues. Remember that the carbon market creates a new source of global wealth. It can use this wealth to close the income divide between rich and poor countries, something that no other international trading agreement has ever been able to do. By giving developing countries more rights to use the global atmosphere than rich countries, and allowing developing countries to sell those rights, the carbon market transfers wealth from North to South, when the CDM applies, while reducing global carbon emissions. It is fair but also productive and efficient. Since China is a poor nation but also the world's largest emitter, it is worth considering how to reach an agreement that caps Chinese emissions or emissions from other poor nations who are large emitters. The rest of this book will address this issue. New and proven technology can help obtain equitable solutions that are effective to avert catastrophic climate change.

7

The Paris Agreement: Failure as an Opportunity

There is a unique opportunity that emerges as a consequence of climate change. The Kyoto Protocol has the potential to unite business and environmental interests, the young and the old. There is an opportunity to forge cooperation between rich and poor, induce much-needed investment in green development, and close the global wealth divide. Put like this, the future looks promising.

Do humans only consider change when we arrive at the point of no return? If so, a new path begins. The time has come in which the promising changes we have described can occur. Business and environmentalist interests can converge, the carbon market can provide the incentives needed for new technologies and we can help pay the cost of the war on climate change. The Clean Development Mechanism (CDM) financing will help to even up the score between the super-rich and those who are in the midst of abject poverty. It is also, however, a race against time.

The 2008–2009 turmoil in global financial markets once again temporarily destroyed faith in markets. In all fairness, this is not new. It is a recurrent theme in a capitalist society. Capitalism is all about taking risks. The entire system is built on the corporation, a unique type of animal that is idiosyncratic to capitalism. The corporation is nothing but a risk-taking machine, an incentive to take economic risks. The distinguishing feature of the corporation is that it caps the losses that it may incur on the downside. This is specific to corporations and is called "limited liability." Through bankruptcy laws the corporation can wipe the slate clean, taking much of the sting out of economic failure and moving to a better future. Since innovation is achieved by trial and error, capitalism encourages innovation. It is no wonder capitalist societies are innovative, creative, and at the same time intrinsically risky. They are built to be this way. And it is no wonder that every so often they get into serious trouble. This is what risk means. It means that most of the time things go well or even very well, but sometimes things go wrong, badly wrong. When that occurs we have the opportunity to improve them, because failure is an opportunity.

The world today is so interconnected that the economic shocks from excesses are transmitted through the world economy with lightning speed. The shockwaves leave no stone unturned.[1] We are moving into a new economic paradigm and the connection between financial institutions the world over is so extreme that any single default can lead to widespread defaults and financial failure.[2]

This also happened in the aftershock of the property market downturn in the U.S. and elsewhere in 2008. Defaults on home mortgages doubled historical standards. The aggregate default rate climbed to about 10% of mortgages at the peak of the housing downturn, creating deep losses in mortgage-based securities that were made by bundling millions of mortgages together.[3] This allowed uncovered positions with huge leverages by the large financial institutions that traded credit default swaps and options over and above mortgage-backed securities, markets that involved over $530 trillion in trades; at the time, almost 10 times the entire economic product of the planet.[4]

Yet as difficult as the situation was for many, it is already yesterday's news for others. The main problems we face are most often caused by the excesses and follies of capitalism. Overgrazing causes desertification just as reckless desire for gains brings us short term profit for some followed by shocking damages for others. It is a cyclical phenomenon. Any economic down cycle will take its course, and will eventually reverse itself. Climate change is different. It can, and probably will be, irreversible.

Investing in a Time of Economic Uncertainty

The larger problem we face now is created by changes that we are unleashing on the metabolism of the planet itself. For the first time in recorded history, humans are changing the planet in ways that endanger its basic life support systems. This is real. Though it could be reversed in geological time, we hope that it is reversible in a timescale that matters for us humans. It is touch and go.

With every challenge comes opportunity. The Kyoto Protocol, the first international agreement based on the creation of a new international market for a global public good, shows the way. If we can implement all solutions through the Kyoto Protocol, we could prevent irreversible losses. New technologies and new market mechanisms operated with the intention of harmonizing our planetary consciousness could create a new start.

As far as the new technologies are concerned, there is a short-term strategy and a long-term strategy, and the one naturally blends into the other. For example, in the short term, carbon capture technologies can relatively quickly capture and safely store away the carbon in the atmosphere, so as to avert climate change. This could be achieved in the next 20 years, using the negative carbon technologies previously discussed. And this strategy can mesh with a long-term strategy of turning away from fossil fuels and towards renewable and carbon free energy sources.

Wind, solar, hydroelectric, and nuclear energies are only our first steps to replace gas, coal, and petroleum. It is now possible to resolve the problem of climate change. We need to use carbon negative technologies as indicated in the Fifth Assessment Report of the Intergovernmental Panel on Climate Change (IPCC). Some of these technologies are available and are becoming a commercial reality right now.

How can we apply the Kyoto magic to other global environmental problems, such as biodiversity and ecosystem destruction? This is not just to raise the bar higher; it is also a practical issue. Without finding a solution to the problem of ecosystem destruction and biodiversity extinction, there may be no human species and no carbon emissions to worry about. We are, after all, in the midst of the sixth largest extinction in the planet, according to scientists, only comparable with the disappearance of the dinosaurs 60 million years ago. Humans would not have survived that period of extinction any more than the mighty dinosaurs did then. It is a serious question whether we can survive this one. We need to understand how a Kyoto-like template can deal with the other global environmental challenges of our time.

The most effective path forward on climate today is direct investment in research, development and demonstration plants of non-polluting energy technologies, as well as negative carbon technologies as described in this book.[5]

It has become obvious that large corporations are inadequate for innovation when it comes to out of the box thinking. The audacious character required to think out of the box is not part of their corporate culture. Successful large corporations are often self-satisfied and avoid risk. Equally, the education and training imposed on students graduating from the "Best Schools" can encourage uniformity and limited originality.

Solving Global Environmental Problems

What about the future of the global environment as a whole? Is there a template to help other fundamental environmental problems of our times, beyond global warming? The short answer is yes, but it requires a bit more explanation as to the why and how.

Human civilization could not have arrived at this point of flirting with disaster, of irrevocably damaging our planet's climate system, without endangering other natural systems as well. The atmosphere's capacity to absorb and to recycle human waste is not the only ecological limit that we have breached. Indeed, this is the reality of the 21st century: just about every natural system worldwide is in a state of serious decline. If the 20th century is considered the century of technological innovation, globalization and knowledge advancement, the 21st century will be considered the century of reckoning — of reconciling our knowledge and capabilities with the earth's delicate balance and our wealth and power with the basic needs of the earth's majority.

From our forests to our fisheries, our atmosphere to our soil, our watersheds to our farmlands, decades of human exploitation and misuse have reached epidemic proportions. We are witnessing a rate of species loss worldwide that is 1,000 times greater than any previous extinction on earth — including the sudden and massive die-off of the dinosaurs.[6] We cannot deny this reality.

Kyoto and the Wealth of Nations

Unless we can arrest the deterioration of ecosystem services worldwide, human wellbeing and prosperity will undoubtedly decline. The links are clear and well understood. Biodiversity loss undermines the basic needs of human societies. For many of us, it is clear that we have arrived at a crucial crossroads in history.

Water is a critical example. The biodiversity in watersheds (ecosystems that encompass our water resources) provides freshwater supplies for drinking and irrigation, erosion control, and purification. These services are essential to human existence and survival. Poorly managed watersheds are unable to provide water filtration or erosion control, which is essential for water catchment and regulating flood waters. Consequently, we now face a global water crisis with profound implications for global food security, human health, and aquatic ecosystems. Population growth, rising standards of living, irrigation for agriculture, and industrial production are pushing freshwater demands to unprecedented levels, while mismanagement, pollution, and climate change threaten existing freshwater supplies. Like climate change, the worst ramifications of the global water crisis are felt by the world's poorest inhabitants.[7] The proof that even the rich and middle classes shall not be spared, however, is California. Television anchors have warned that pools and gardens have to go on a diet. The inaction demonstrated by the governing institutions in harvesting water is proof either of ignorance or incompetence. Simple techniques of water harvesting have been around for hundreds and thousands of years and were known to the Romans and many other cultures.

The climate crisis is the epitome of the global ecological crisis and the future of the planet and of humankind is at stake. The vitality of ecosystems, basic human needs, and equity are intertwined. But from this common ground we can build anew.

The strategy for the future is not only to replace what has been lost — if that is even possible — but to invest in what still remains.

The lesson we have learned is that we cannot easily and cheaply replicate ecosystem services with human-engineered substitutes. For example, human-built water filtration systems require billions of dollars to replace the natural filtration services that biodiversity provides at no cost. Viewed in this light, the degradation of watersheds makes little economic sense, whereas investing in watershed restoration and protection does.[8]

In 1996, New York City confronted two options for bringing its water supply in line with Environmental Protection Agency (EPA) regulations: it could invest in natural capital or physical capital. Which did it choose? New York City's water comes from a watershed in the Catskill Mountains. The Catskill watershed was so degraded by fertilizers, pesticides, and sewage that it could no longer filter and purify the city's drinking water to meet EPA standards. Investing in natural capital meant buying critical lands in the watershed, restoring, and protecting these lands and subsidizing the construction of better sewage treatment facilities. The total cost for restoring the watershed was estimated to cost between $1–1.5 billion.[9] What about the alternative?[10]

The alternative was to build an artificial filtration plant with sufficient capacity to clean New York City's water supply. The estimated cost for this capital project was $6–8 billion. The choice in this case was straightforward, and by investing $1.5 billion in natural capital, the city saved $8 billion in physical capital costs. The bottom line was better than these estimates led city planners to believe. By restoring the watershed, the city helped to protect other important ecosystem services, such as biodiversity and carbon sequestration.[11]

We face similar choices now with respect to our forest resources. Plantation forests can stabilize soil loss and sequester carbon, but they cannot replace the diversity and vitality of natural forest ecosystems. Yet forests are disappearing worldwide at an extraordinary pace. We lost an average of 5.2 million hectares of forests each year from 2000–2010, an area the size of Panama or Sierra Leone.[12] At present, the only way to realize the value of our forests is to destroy them, to sell the wood for pulp or to burn it to give way to arable land. The attendant ecosystem destruction undermines the health needs of populations around the world, in rich and poor countries alike.

Environmental degradation worldwide results in a loss of ecosystem services estimated well over $2–4.7 trillion each year.[13] We lose almost the equivalent amount in natural capital each year as what the dot-com bubble lost in the early 2000s, wiping out an estimated $5 trillion in wealth.[14] And it is natural capital, not financial capital, which ultimately determines human wellbeing and even survival.

The air we breathe, the water we drink, the soil from which we grow our food, the fuel that powers our production, the traditional knowledge that informs our medical techniques, the species that enrich our psychic and emotional health, and the amenities we rely on for recreation and our enjoyment of natural beauty; these are what comprise the planet's natural capital stock. And this stock of wealth faces much graver dangers than the runaway lending, greed, and plummeting home values that threatened the global financial system. Once this wealth is gone, it cannot be recreated. We must find ways of protecting it.

There is an enormous disconnect worldwide between the value that biodiversity provides to humans, which derives from satisfying basic human needs for water, food, and medicines, and the economic value that can be realized from biodiversity in terms of dollars and cents. The ecosystem services we depend on are almost all examples of public goods. Like the atmosphere, we can trace the degradation of these critical ecosystem services to the absence of prices and clear rules of ownership. These are resources that all people on the planet share or have access to. We have used them and abused them throughout history, with little consideration for how our use diminishes the value of these resources for others or for future generations. We have never had the right incentives to conserve these resources because there was never the right price tag attached to their use. As a result, we may now pay the ultimate price.

Biological diversity, the planet's store of wealth and of knowledge, belongs to all of us and to none of us at the same time. It belongs to the future. That future depends on investing in the natural capital that is abundant on the planet and harnessing its potential for the betterment of all. The earth and all of its natural riches cannot be sold today to the highest bidder.

The Solution?

In the appropriate context, ecotourism is an excellent case-in-point. Ecotourism has emerged as the primary source of foreign exchange earnings for countries such as Costa Rica, Guatemala, and Thailand. Foreigners from wealthier but environmentally poorer areas pay to experience the rare biodiversity within destination countries. These countries are, in essence, selling access to their resources but making sure that the underlying assets, the forests, and the biological diversity, are not for sale. Ecotourism allows the host country to generate a stream of much-needed revenue from its stock of natural wealth. In so doing, it provides an incentive to conserve and protect ecosystem wealth rather than to liquidate it in order to sell the attendant forest products and agricultural commodities in international markets.

Within limits, ecotourism is a service that some developing countries are uniquely positioned to sell because of their abundance of natural resources. In this sense, it is potentially an equalizing force between rich and poor, since most remaining natural resources are in poor nations, because rich nations have used most of their biodiversity to industrialize. The primary concerns with ecotourism, however, are the cultural and social impacts that it can produce, and how the dividends from that wealth are distributed within the host country. It is critical that the host country distributes the revenues from ecotourism fairly; otherwise its citizens will be denied their share of the benefits from preserving their natural habitats and will lack incentives to protect it. These issues determine the long-term success of ecotourism or of any global environmental market, in promoting conservation. Much of this depends, however, on a country's own internal wealth and income distribution, and cannot be solved at the international level. It is important to note that the same is true for climate change or any other global environmental issue. The UNFCCC can distribute the burden of emissions reduction fairly between countries, but individual countries are the only ones empowered to decide how to distribute the costs and benefits across their own populations.

Most developing nations have an abundance of natural resources: forests, minerals, fossil fuels, fisheries and the attendant biodiversity. With so much wealth, how is it that these countries struggle to satisfy basic needs? It is because the global economy, as currently structured, has no mechanisms for recognizing and rewarding natural wealth. For the moment, the only way for countries to capitalize on this wealth is to destroy it, selling products made from these resources in international commodity markets, with little additional value; for example, selling raw wood, rather than furniture, and coffee beans rather than coffee. Cutting trees to reproduce toilet paper is an excellent example. A standing tree has no market value today while toilet paper has a well-defined market value. The lack of property rights leads to undervalued resources and ensures that developing countries are paid as little as possible for the goods they exports, while paying top dollar for the finished goods and services they import. Reliance on natural resources for export condemns developing nations to poverty while destroying the planet's rich biodiversity. There is a name for this phenomenon: the natural resource curse. It plagues some of the most biologically rich and diverse countries in the world. With the carbon market, we could turn this curse into fortune.

Markets for global public goods, have unique properties that link equity and efficiency in ways that stand apart from all other markets, and they may help solve this dilemma. The world is waking up to the need to protect our remaining natural capital. Much of the demand for environmental protection comes from rich countries, where centuries of industrialization have left them with a debit in their environmental account. Developing countries are in the opposite situation: they have a credit in the environmental account but a financial deficit. They have produced less and they house most of the world's remaining forests and biodiversity. There are, therefore, natural gains from trade between the two groups of nations.

Global biodiversity, like emissions abatement, is a global public good. Once biodiversity is protected, all countries benefit from it, albeit to different extents. Countries cannot choose different levels of global biodiversity based on their incomes and preferences. This is what makes a market for global public goods different from all other markets. To bring each country's demand — its purchasing power — in line with the level of biodiversity protection that the global community has set as its goal, we will have to shift wealth from rich to poor countries. To do so is not only fair, but it is a condition for market efficiency in markets for global public goods.

Right now, rich nations may wish that developing countries achieve a higher level of biodiversity conservation than what they might otherwise choose, given their pressing development needs. This is similar to how the Kyoto Protocol requires developing countries to sign up to a higher level of global emissions reduction than they may presently afford. Just as it made sense and was efficient for the Kyoto Protocol to allocate emissions rights to the benefit of developing countries, so it is fair and efficient to have payments for biodiversity conservation flow to developing nations.

Markets for global public goods facilitate payments for biodiversity conservation. Markets for global public goods cannot emulate, in theory or in practice, any other markets that have existed so far. Because they trade in a good that is universally consumed, such as the stability of the atmosphere and biological diversity, efficiency demands that the market works to the benefit of poor nations. Done right, payments for biodiversity conservation can be an equalizing force between nations: a way of leveling the playing field between rich and poor in an era of growing environmental awareness and resource needs.

To create new global markets for public goods, however, we have to decide who owns the natural wealth. Markets cannot trade in goods and services whose ownership is not clearly defined. We need to define this wealth as common wealth, and distribute user rights to the countries who need them most. This is what the Kyoto Protocol achieved with respect to the atmosphere. It created user rights to the global atmospheric commons and distributed those user rights to the benefit of developing nations.

Without the Kyoto Protocol, emitters effectively claimed the planet's wealth, our common wealth, for free, and used it for their own private gain. By emitting greenhouse gases into the atmosphere, they diminished the atmosphere's climate regulating capacity. In a similar way, the international oil companies, the multinational logging interests, the bio-prospecting pharmaceutical companies, the agricultural behemoths and even rich nations themselves have pillaged the natural wealth stock of the planet for decades. The time is right to reclaim this wealth as our common asset and to share the dividends reasonably between us.

The Kyoto Protocol Shows the Way

Payments for biodiversity conservation could, in principle, provide incentives for preserving and investing in the planet's natural wealth, and are similar to the voluntary carbon markets that were created in the U.S., for example at the Chicago Climate Exchange. To achieve this goal in a sustainable fashion:

- They must promote the conservation of biological diversity.
- They must encourage sustainable use and equitable access, and benefit the sharing of biological resources across the world.
- The payments should be financially self-sustaining and incorporate local communities, governments, and the private sector.
- The payments must address the basic needs of developing countries, especially those nations' poor, their women, as well as their indigenous and local communities.

These principles are a tall order. Are payments for biodiversity conservation up to the task? While they have achieved some successes, it is generally recognized that, without a proper market structure including assigned property rights, payments for ecosystem services and other voluntary markets do not have a future. No other financial mechanism or market is subject to such demanding conditions as the four listed above except, for the Kyoto Protocol carbon market, which satisfies them all. Then again, no other market mechanism trades directly in the wealth of the planet.

The carbon market is the most developed form of payment for an ecosystem service operating at the international level. Through Kyoto's CDM, projects in developing countries that deliver legitimate and certifiable carbon offtakes receive payments from carbon emitters in developed countries. The bad guys pay the good guys; the rich countries pay the poor. The carbon market is fair and because of the unique properties of global public goods, it is also efficient.

The UNFCCC experiment with innovative and positive incentives for combating climate change, provides inspiration for addressing other global environmental issues. It supplies a model for combining equity and efficiency in protection of the earth's endangered biodiversity. Thus, there is more at stake in saving the Kyoto Protocol than averting climate change.

The U.N. carbon market provides:

- A carbon price signal that rewards carbon removal and penalizes excessive emitters, thus helping to avert global warming.
- A means to fund emissions reduction without donation or the need to create global taxing authorities. It is self-funded.

- A means to transfer wealth from rich to poor nations in the form of productive and clean investment through the CDM.

Can we emulate the potential of the Kyoto Protocol in other areas where ecological crises now manifest? Yes, but only if we can save the Kyoto Protocol. If the global community can solve the climate crisis, we can solve other global environmental problems such as water scarcity and biodiversity destruction. It is possible.

Thoughts and Possibilities

For the first time in human history, the future of human societies hangs in the balance. Developing nations are in a position to tip that balance one way or the other. They are home to the large majority of the world's population, forests, biodiversity, languages, cultures and indigenous knowledge. The future depends on whether we can preserve these precious resources and overcome the global divide.[15] We know that business-as-usual is a recipe for their destruction. If developing nations pursue the same development path that rich nations set out on two centuries ago, the world will be a far more diminished place as a result. Rich countries blazed a path that was as ruinous as it was profitable. The path led to phenomenal wealth and prosperity for a minority of the world's people and ecological destruction for the planet. How can we close that path to the majority, without preventing access to prosperity? Are there other pathways to the future?

An enormous dilemma has been created: a conflict between rich and poor over access to resources defines the geopolitics of the 21st century. How will this play out? Rich corporations can use their power and wealth to control the world's endangered resources, or they can work cooperatively with developing countries to chart a cleaner development course. The rich can try to preserve their current lifestyles while condemning the developing world to poverty, or they can reduce their own demands on the biosphere to allow room for developing countries to preserve their real wealth: natural ecosystems. The defining conflicts in the 21st century will be over access to the world's remaining resources.

Much of the U.S.'s antagonism toward the Kyoto Protocol can be understood in this context. The U.S. is not yet willing to concede the need to alter its own course, yet it understands the implications if other countries emulate its growth. It finds it easier to point a finger at China and its growing energy needs, or India with its many mouths to feed, than to confront its own resource overuse.

No matter how hard rich nations may try to sustain current trajectories, it simply won't work. Even if rich countries denied the entire developing world access to the atmosphere, the emissions from the industrialized world's own production and consumption would be sufficient to unleash climate change.

The pathway to the future requires an energy revolution in the developed world. Fossil fuels are inextricably linked to our energy system, and it seems impossible to eliminate their use overnight. Energy efficiency is a large and mostly untapped resource, but even this cannot take us all the way to where we need to go. It is difficult to drastically decrease the world's use of fossil fuels in the immediate future, yet this is what we need to do. This explains the interest in new technologies for capturing carbon, either as emitted by fossil fuels plants or directly from the atmosphere, and storing it safely.

Technology, however, cannot act as a substitute for taking the necessary next steps toward weaning our energy systems off fossil fuels and reducing emissions. To do this, we need a strong price signal — the cost of producing carbon emissions must provide the right incentives to minimize fossil fuel use. For industrialized countries the price signal comes from the cap the Kyoto Protocol imposes on their carbon emissions. For developing countries the price signal comes from Kyoto's CDM and the opportunity cost they incur for producing emissions rather than carbon off-takes. The benefit of technologies for capturing and storing carbon is that they do not interfere with the price signal. For example, in the U.S., there is increasing pressure to resume oil drilling in protected areas in Alaska, to ease the burden of higher energy prices until alternatives to oil can be discovered. This contradicts the rationale of climate policy and takes us further from the goal of emissions reduction. By increasing the supply of oil in the hope of lowering its price, it would undermine incentives for greater energy efficiency and investment in renewables.[16] Drilling is a stopgap measure for lowering energy prices, not for building a stable climate future.

Building a new energy infrastructure, capable of delivering power from renewable and clean energy sources to our homes and our factories, could take time — time we don't necessarily have. If the global community had acted sooner when evidence of climate change emerged, we might have more options now for dealing with the crisis. Waiting for new technologies to miraculously appear in the absence of a strong price signal may be unproductive. Assuming annual improvements in energy use and energy technologies lead us to a paradoxical conclusion. Because climate change is a long-term crisis and predictable, inexorable technological change will make it easier and cheaper to reduce emissions in the future, it appears better to wait before addressing the problem.[17] This view of technological change leads to very cautious recommendations about how low to set our emissions caps in the near future. Waiting may seem to be a better option to those who believe that the problem can solve itself in time. But waiting is exactly what we cannot risk doing.

Getting the price right for carbon is the first step in combating climate change. This is why we cannot afford to let the Kyoto Protocol and its carbon market disappear by removing its firm caps on global emission, which can strengthen and prolong the price signal. We can only arrive in the future with new energy technologies if we start the conscious, carefully planned development of those technologies today. Waiting for the *deus ex machina* of technological change will ensure that we face fewer options in the future at dramatically higher costs.[18]

Solving the climate crisis requires concerted investments in research and development. We need the equivalent of a Manhattan Project (the joint U.S., U.K., and Canadian project to develop the first nuclear weapon during WWII) for the new energy technologies of the 21st century. And we need a New Deal — a Green New Deal — for building the infrastructure to support it. As recent studies have shown, recycling revenues from the sale of carbon allowances in national carbon markets can actually stimulate economic growth, in part by redistributing income from rich to poor — from those who save to those who spend — and creating new jobs.[19] The price signal can work its magic by offering incentives to the private sector to develop the industries of the future, and the CDM can deliver these advances to the developing world. But few countries will embark on this ambitious endeavor alone, unless their efforts are nested in a larger global effort to solve climate change.

New technologies can stimulate new investment, save consumers money, spur productive research and development with spill-over benefits for other sectors, create new jobs and help to reduce energy imports while increasing technology exports. Massive public investment in military technology since WWII led to the widespread adoption of jet aircraft, semi-conductors and the Internet by industries and households, and is responsible for the technological advantage the U.S. holds globally. If the rest of the world moves forward with the Kyoto Protocol without the U.S., the U.S. risks losing its technological advantage globally, unless it charts (and funds) a careful and deliberate new technological path.[20]

Countries such as the U.S. must commit to solving climate change in a precarious financial environment. Years of irresponsible deregulation of global financial markets, stoked by the very mindset that encourages a self-regulating, cautious, business-as-usual approach to climate change, has almost brought the global economy to its knees. This was a potentially ominous new sign for the climate negotiations in Copenhagen in 2009; the Europeans were already showing weakness in their commitment to Kyoto and its carbon market. Paris COP21 created an even shakier ground since the Paris Agreement has no mandatory emissions limits. In the U.S, President Barack Obama passed crucial legislation limiting emissions from vehicles and from stationary sources such as power plants. However, the Trump administration is returning to the past.

The conditions are now ripe for a leap forward. It had been hard to convince people to make changes in their lives when by all outward appearances, everything was fine. This has always been the difficulty: convincing people of the urgency of the climate crisis when its worst effects were still far off in the future. But climate change is happening now, and is here today; global consciousness about climate change is rising, and almost everyone recognizes the need for change. Massive natural disasters — such as the hurricanes, typhoons, floods, and wildfires that increased rapidly in frequency and scale due to additional energy in the atmosphere and in water bodies — are the signal.

The magic of Kyoto is that it can avert the risk of global warming at no net cost to the global economy. Its carbon market accounts for the cost of emissions to the global environment by using an efficient free market mechanism. And Kyoto does all this without requiring any donations to support its own performance. On the contrary, its carbon market creates a new source of funding for projects that promote global energy and wealth, which will help close the wealth gap between rich and poor nations.

The Kyoto Protocol obviously has the potential to bring the peoples of the world together. It carries the possibility of promoting global ecological restoration and financing the war on climate change. Yet, the situation now is more uncertain and potentially catastrophic than ever.

8

Avoiding Extinction

What Next?

For the first time ever, humans dominate planet Earth. We are changing the basic metabolism of the planet: the composition of gases in the atmosphere, its bodies of water, and the complex web of species that makes life on Earth. What comes next?

The changes we are precipitating in the atmosphere are fundamental and can lead to disruptions in climate and global warming. Both the North and the South Poles are melting. Water expands when it is heated. Since the seas are warming, they are rising all over the world. This irrevocable upward trend is well documented: slowly but surely the rising waters will sink most island states — there are 43 island states in the United Nations, representing about 23% of the global vote, and most or all could disappear soon under the warming seas.

The current shift in climate patterns has threatened many species. It has also allowed for the spread of certain species; some insects are migrating to areas they did not previously inhabit, bringing with them a variety of vector-borne illnesses. For example, new outbreaks of malaria in Africa are on the rise. Humans are also shifting ground. The U.N. reports that more than 25 million people are migrating due to drought and other climate change-induced conditions, and the numbers are increasing rapidly. The 2014 migration of over a million people to the EU caused considerable political stress and led to anti-immigration candidates in German, the U.K., and the U.S. This is just the beginning.

In the U.S., the physical consequences of climate are less extreme but still evident. The mighty Colorado River is drying up, prompting orders to turn off farm water. Miami has the largest risk among U.S. cities with damages anticipated in the trillion of U.S. dollars, and it already suffered record losses in 2017. Lake Mead's waters in Nevada are exhibiting record lows, threatening the main supply of water to Las Vegas. Wildifres from drought conditions have multiplied and spread rapidly in California, reaching their highest levels since 2006.

The world is aware of the connection scientists have found between climate change and the use of fossil energy. The largest segment of carbon emissions, 45% of global CO_2 totals originates in the world's power plant infrastructure, 87% of which consist of fossil fuel plants that produce the overwhelming majority of the world's electricity. This power plant infrastructure represents $45 trillion according to the International Energy Agency (IEA), almost the scope of world's economic output. New forms of clean energy are emerging such as wind farms in Scotland and solar plants in Spain and the U.S. in an attempt to forestall carbon emissions. But the process is necessarily slow. Since the world's power plant infrastructure is comparable in monetary value to the world's entire GDP, changing this infrastructure can take decades. Transforming the power plant infrastructure is too slow to avert the potential catastrophes anticipated in the next 10–20 years. What is the solution?

We propose a realistic plan that involves market solutions in industrial and developing nations, simultaneously resolving the problems of economic development, climate change and conflict in global climate negotiations. Climate change is just one of several global environmental areas that are in crisis today. Biodiversity is another; industrialization and climate change threaten the world's ecosystems. Endangered species include sea-mammals, birds such as cockatoos, polar bears, tigers, and marine life such as coral, sawfish, whales, sharks, dogfish, sea turtles, skates, grouper, seals, rays, bass, elephants, and even primates, our cousins in evolution. Scientists know that we are in the midst of the sixth largest extinction of biodiversity in the history of our planet, and that the scope of extinction is so large that 75% of all known species are at risk today. The U.N. Millennium Report documents rates of extinction 1,000 times higher than fossil records. The current extinction event is the largest since the dinosaurs' extinction 60 to 65 million years ago. But today's extinction event is unique in that it is caused by human activity. And it puts our own species at risk. There is a warning signal worth bringing up: all major recorded planetary extinctions were related to changes in climate conditions. Through industrialization we have created environmental conditions that could threaten our own species' survival.

We know that 99.9% of all species that ever existed are now extinct.

Are we next?

Will humans survive?

The issue now is how to avoid extinction.

Bacterial Altruism

To avoid extinction, we have to develop survival skills for a changing environment. This seems reasonable and natural — yet the social skills we need are not obvious. These could be quite different from what human societies have achieved so far, the individual survival skills that we are familiar with. A simple but somewhat unexpected experimental finding involves colonies of bacteria, which are one of the world's oldest living species. They have been around for billions of years and have shaped the planet's geology and its atmosphere to suit their needs. Bacteria are champions of survival. They needed appropriate survival skills, and have developed unexpected skills based on what can be described as "altruism." Since bacteria are among the oldest species on the planet, much older than relatively recent humanoids, we need to take their skills seriously as a model of survival. Bacterial colonies know how to avoid extinction. New findings indicate that *Escherichia Coli*, and indeed most known bacteria colonies, when exposed to a pathogen or stressor such as antibiotics, not only mutate and evolve to develop resistance, but also produce specific resistance tools that evolved members do not need, in order to share with the rest of the (non-evolved) members of the colony.[1] In other words — when exposed to stress, mutant bacteria use some of their own energy — altruistically — to create a chemical called "*indole*" that protects non-mutants from the pathogen. This way the entire group survives. A way to summarize this finding is to say that *altruism* is an effective survival tool and that bacteria — those champions of survival — have developed and mastered altruism for this task.

This finding is quite different from what we believe to be effective survival skills in human colonies or societies. Until now, human survival skills have focused on avoiding natural risks and confronting successfully the threats posed by other species that preyed on us, species that are dangerous to us. Altruism to a certain extent has been considered almost a weakness in human societies; it is considered to be a desirable ethical trait rather than a survival skill. Yet, it is a survival skill. Aggressive and individualistic behavior may have been a useful survival tool until now. The war society that humans have created has become an efficient killing machine. But when things change, as they are changing right now, strengths can become weaknesses. Things have fundamentally changed and they continue to evolve quickly. Physical strength and aggression matter less today for human survival than intelligence. Some of the worst risks we face today are caused not by other species that prey on us, but by traits that we evolved because they succeed against our predators — for example, extracting energy and burning fossil fuels in order to dominate nature and other species. In other words, we are now at risk due to human dominance of the planet. Our success as a species has become the source of our main risks. Humans are causing some of the worst risks humans face. The situation is somewhat unusual and new for our species, and it is also new for the planet itself. As the situation changes, the rules we used to follow for survival must change too.

Let us start from basic principles. Survival is about protecting life, not just about inducing death. Life is difficult to define, but we all agree that it is a phenomenon characterized by reproduction. Only those systems that incorporate reproduction are said to be alive. Life forms are able to reproduce. To be alive means to be part of a series of reproductive activities. Reproduction characterizes life. Destruction does not. Asteroids destroy very effectively, and so do volcanoes. But they are not alive, because they do not reproduce. Humans are alive because we do.

But reproduction fundamentally requires altruism rather than dominance and aggression. How so? This is simple. We must donate our energy, and even our bodily resources and substance, to be able to reproduce.

In our culture, the essence of survival is viewed differently. It is generally viewed as the ability to conquer, dominate, and kill. Research shows that men tend to think of life skills as those skills that allow them to win the battle for survival. War is an example. Surveys asking men what characterizes life, find that they are likely to say "the survival of the fittest" or "dog eat dog." This may be because of the evolutionary role that males originally had in human societies, a role that is somewhat outdated. The reality is that humans could not live, nor indeed be part of the chain of life, if we did not have the nurturing skills needed to reproduce. Women understand that reproduction requires altruism. Women donate their physical substance, such as eggs, blood, and milk, and they do so voluntarily for the sake of reproduction. This is what reproduction is all about: voluntary donation of one's substance. Most living beings, animals and plants, do the same. They donate their substance voluntarily to the next generation, sometimes at the cost of their own welfare and even their own lives. Voluntarily donating one's own substance, flesh and body fluids, is the very essence of altruism. This altruistic donation is key to the survival of the species. The great British author and social commentator Jonathan Swift once suggested, as a "humble proposal" to resolve the problem of hunger in Ireland that humans should eat their own children. This is not as outlandish a proposal as it may sound at first. In any case it helps to illustrate the point. If the essence of life was the survival of the fittest, then humans would eat their children who are powerless at birth — nobody is less fit than newborn infants. Their bodies could certainly provide protein and nutrition to fit adults.

The question that we must answer is: Why don't we follow Swift's humble proposal? Why not eat our own children?

Some societies may have done exactly that, but those societies are not here to tell their tale just as if we ate our children, we would no longer be around.

Our species would not have survived. No species that eats its children would survive — it may not even get started as a species. Except for aberrations, survival depends crucially on reproduction and this means protecting the weak — including the weakest of all — small children. This is quite different from the blanket policy of survival of the fittest, with regards to the adult members of the species. Indeed, one may say that survival is more than anything about altruism and cooperation, and about the protection of the weakest. It is not about "dog eat dog" — it is not about dominance and survival of the fittest. It is about the nurturing and protection of new generations; it is about voluntary donations, about the protection and nurturing of the weakest, sometimes at the expense of our own survival.

Women and Survival

Women's evolutionary role is to protect the weakest of all: children at birth. Women are critical to human survival since they are the key to reproduction, voluntarily providing their substance and energy to give birth and protect the weakest, as needed for the survival of the species. Men miss this important aspect of survival because their evolutionary roles appear to value physical strength more than anything else. However, this is a role that seems increasingly out of date.

It is fitting to raise the issue of "avoiding extinction" within a male-dominated world and culture that is focused on violence, economic competition, and wars of choice. We need to assure a changing role for women, so the entire ethos of destruction and dominance that permeates our society is balanced out by a modicum of altruism. Nurturing and protecting the weakest is critical and necessary if we are to avoid extinction.

It is true that there have been changes in the role of women, most of all with their rapid entrance into the market for labor in industrial societies. But this change has not changed matters. Modern societies, such as the U.S., still witness abuse of women at home and elsewhere, both physically and economically. The U.S. Congress created a committee to investigate violence against women. The U.S. has a 20% gender difference in salaries, which does not seem to be narrowing. There are salary differentials even when controlling for equal training, age and experience. Gender inequality is prevailing, persistent, and systematic. In any given society, there is a statistical correlation between the amount of housework a woman does at home and the difference between male and female salaries in the economy as a whole. These two different statistics — two indices of abuse — are indeed related, because when women are overworked and underpaid at home, this leads women to be overworked and underpaid in the marketplace.[2] Gender inequality in salaries is, in reality, legally sanctioned: the U.S. still does not have an Equal Pay Act, so unequal pay for women and men is still legal in the U.S.[3]

Is there a reason to pay women less than men? If so, what is it?

These persistently unequal situation is based on a rationale of "genetic inferiority" in women. Even a former president of the oldest University in the U.S., Larry Summers from Harvard University, presented this suspicion in public as a plausible hypothesis to explain the persistent >20% difference in salaries between women and men in our economy. When he was subsequently voted out by Harvard University faculty, he went on to become an economic advisor to then President Barack Obama. One wonders whether Mr. Summers would have been selected as an economic advisor to the president of the U.S.— the first black U.S. president — if he had presented in public his suspicions about the genetic inferiority of blacks, rather than the genetic inferiority of women. I venture to say he would not have been selected by President Barack Obama if he had said in public that blacks are genetically inferior. But saying this about women is acceptable, and he went through and was rewarded by President Obama with an economic advisory role. This was a discouraging event for many, but not for the men who secretly or openly believe that women are indeed genetically inferior to men. One cannot but draw a somewhat distant but illustrative connection between this situation and the excuses used to explain anti-semitism based on the supposed genetic inferiority of Jews. This illustrates the implications of claiming genetic inferiority for some groups in our society.

Publicly declaring the genetic inferiority of women to explain their economic exploitation is not an innocent remark even if the genetic inferiority at stake is about performance in the sciences. It is a way to justify a systematic method in which male-dominated societies perpetrate economic and cultural abuse, violence and brutality against women, pornography, torture of women, and rape, which often represents a form of social control and intimidation. Ultimately it is a deep social rejection of altruism — the protection of the weak and the essential reproductive role that women bring to society — which is a necessary precondition for the survival of the human species. Our society's manifested hate and violence against women — misogyny — is critically connected with the self-destructive aspects of our society — and the problem of avoiding extinction that we face now.

Until we change the current male-dominated culture and its abuse and barbaric treatment of women — for example, until we revolt against the acceptance of electronic games involving the systematic torture and killing of women as entertainment, which the U.S. Supreme Court found acceptable for children in a 2011 decision — and until we develop altruism as an efficient survival skill, our society will not be well prepared to avoid extinction.

Avoiding Extinction: Summary of What is to Come

The future of humankind may be played out in the rest of this 21st century. Here is a summary of the situation and what to do about it —which is developed further in this chapter.

In a nutshell, environmental resource limits confront us with massive scarcity today and in the future. The problem of overuse of natural resources, continues to pose a clash of civilizations: it is an impasse between the global North and the global South. The North refers to the rich nations that inhabit mostly the Northern hemisphere of planet Earth, the South the poor. The former represent less than 20% of the world's population, and the latter more than 80%. We will examine the market's role in getting us here and finding a solution, and we will also define building blocks that are needed for a solution going forward: how to bridge the global wealth gap, how to transform capitalism as needed for this purpose, and whether this is all possible. We will examine the role of the United Nations Kyoto Protocol and its carbon market in this global transformation by itself and in conjunction with other global markets for environmental resources — for water and biodiversity — yet to emerge. We have examined the critical role of women, and will now turn to examine how the global financial crisis fits into all this, how it affects our future, and the lessons we have learned.

Avoiding extinction is the ultimate goal of Sustainable Development.

Financial and Global Environmental Crisis

While we are still climbing up from the depths of a global financial crisis that started its deadliest stages in 2008, the world knows that the game is not over. Judging by the threats from the Eurozone including Brexit, it could all re-start next year. For the first time in history, the U.S. was downgraded to a debtor nation a few years ago, and the shocks to its financial markets underscore these points. At the same time, within a larger historical context, the financial crisis takes second place. We have seen such a crisis before. What we have never seen before is the global threat to human survival that is developing in front of our own eyes. We are in the midst of a global environmental crisis that started in a small way with the dawn of industrialization and accelerated with the onset of globalization, ever since the Bretton Woods Institutions were created after WWII to provide a global financial infrastructure for spreading markets and industrialization across the world economy. In both cases, financial mechanisms are at work. The global financial crisis and the environmental crisis are essentially two aspects of the same problem. How so?

Simple examples are available in everyday media. The urgency of the situation has become clear. On June 21, 2011, *The Times* in London wrote that "marine life is facing mass extinction" and it explained: "The effects of overfishing, pollution and climate change are far worse than we thought. The assessment of the International Program on the State of the Oceans (IPSO) suggests that a 'deadly trio' of factors — climate change, pollution, and overfishing — are acting together in ways that exacerbate individual impacts, ... The heath of the oceans is deteriorating far more rapidly than expected. Scientists predict that marine life could be on the brink of mass extinction." *All three causes* of extinction just mentioned — overfishing, pollution, and climate change — are attributable to the industrialized world, which consumes the majority of the marine life used as seafood, generates about half of the global emissions of carbon dioxide and uses 70% of the world's energy. Industrialization is at work, contributing to the impending destruction and mass extinction in the earth's seas. Financial markets are at the heart of the industrial economy.

The complexity of the environmental problems is baffling scientists. The Earth self-regulates its atmosphere, but right now we are tying the Earth's hands in self-regulating itself. There is no quick fix. A standard way that the planet uses to regulate carbon, for example, is to sequester carbon from the atmosphere in its mass of vegetation, which breathes CO_2 and emits oxygen. Animals, such as humans, do exactly the opposite. Animals breathe oxygen and emit CO_2. In balance, the two sets of realms — flora and fauna — maintain a stable mix of CO_2 and oxygen. Since CO_2 in the atmosphere regulates its temperature, this cycle maintains a stable climate. But the enormous use of energy by industrial societies is tipping the scales, and our widespread destruction of the mass of vegetation prevents the planet from adjusting. What about planting trees? Can't they do the job? On the same day, June 21, 2011, *The Times* mentioned that, "planting trees does little to reduce global warming" and explained how a recent Canadian report[4] had found that, "even if we were to plant trees in all the planet's arable land — an impossible scenario with the global population expected to rise to 9 billion this century — it would reduce less than 10 percent of the warming predicted for this century from continued burning of fossil fuels."

Observe that it is not the developing nations that are causing this problem. This is because over 70% of the energy used in the world today is used by the less than 20% of the world population that live in industrial nations, which originate about half of the CO_2 emitted by humans. These are the same industrial nations that created the Bretton Woods Institutions in 1945 and have consumed an overwhelming amount of the Earth's resources since then.[5]

For these reasons, one can say that the financial crisis and the environmental crisis are two sides of the same coin. They are at the foundation of the current model of economic growth in industrial nations and of its voracious use of the Earth's resources. One can pinpoint precisely which part of our economic model destroys the environment and creates financial crisis: it is the practice of "discounting the future" which was introduced by the famous economist Tjalling Koopmans, who gave it the name "impatience." In financial markets, it is also called "short termism" and can lead to Ponzi schemes. When "discounting the future" comes into play in environmental and natural resource issues, we ignore the future needs of the planet and our species. Sustainable development requires an equal treatment of the present and the future, an axiom that I introduced when I defined the formal theory of sustainable development. In a nutshell: both the world's financial crisis and the global environmental crisis stem from a flawed economic mindset, and both require a new model of economic growth that is characterized by sustainable development.

This view is shared by the recently created international group G20, the first leading group of nations that includes developing countries. The group met for the first time in Pittsburgh, U.S., on September 24–25, 2009. The G20 Leader's Statement (September, 2009) says:

As we commit to implement a new, sustainable growth model, we should encourage work on measurement methods so as to better take into account the social and environmental dimensions of economic development. Modernizing the international financial institutions and global development architecture is essential to our efforts to promote global financial stability, foster sustainable development, and lift the lives of the poorest. Increasing clean and renewable energy supplies, improving energy efficiency, and promoting conservation are critical steps to protect our environment, promote sustainable growth and address the threat of climate change. Accelerated adoption of economically sound clean and renewable energy technology and energy efficiency measures diversifies our energy supplies and strengthens our energy security. We commit to: Stimulate investment in clean energy, renewables, and energy efficiency and provide financial and technical support for such projects in developing countries; and Take steps to facilitate the diffusion or transfer of clean energy technology including by conducting joint research and building capacity. The reduction or elimination of barriers to trade and investment in this area are being discussed and should be pursued on a voluntary basis and in appropriate fora.

The G20 statement continues:

Each of our countries mo will need, through its own national policies, to strengthen the ability of our workers to adapt to changing market demands and to benefit from innovation and investments in new technologies, clean energy, environment, health, and infrastructure. It is no longer sufficient to train workers to meet their specific current needs; we should ensure access to training programs that support lifelong skills development and focus on future market needs. Developed countries should support developing countries to build and strengthen their capacities in this area. These steps will help to assure that the gains from new inventions and lifting existing impediments to growth are broadly shared.

...We share the overarching goal to promote a broader prosperity for our people through balanced growth within and across nations; through coherent economic, social, and environmental strategies; and through robust financial systems and effective international collaboration ... We have a responsibility to secure our future through sustainable consumption, production and use of resources that conserve our environment and address the challenge of climate change.

The G20 knows the problems that nations face. What they do not know are the solutions. On April 30, 2016, *The Economist* reported a new measure of economic welfare introduced by James Tobin, the famous Yale economist. A 2009 report commissioned by the former French President Nicolas Sarkozy, chaired by my Columbia colleague Joseph Stiglitz, a prominent economist, called for changes in our measurement of economic progress and growth and for an end to "GDP fetishism" in favor of a "dashboard" of measures that capture human value. These reports offered appropriate criticisms, recognizing the problem at hand. "The report is in part a response to environmentalist concerns that GDP treats the plunder of the planet as something that adds to income" wrote *The Economist*. "The report was much talked about: it was not much acted." Once again, the problem is identified, but solutions are lacking. We turn next to the solutions.

Human Future: Green Capitalism

The task in front of us is nothing less than building a human future. In the midst of the sixth largest extinction on planet Earth, and potentially catastrophic climate change, we face extinction of life on land and in the seas, the basis of life on Earth. We have come so close to the brink in the current generation that it appears right now that only a new different generation can help. As Albert Einstein said: "The mindset that created the problem is not the mindset that will find a solution."

We need to stave off biodiversity extinction and reduce carbon emissions, while rebuilding the world economy and supporting the needs of developing nations. Is this possible?

It is. To understand the solutions, we need to look closer at the root of the problem.

The World since WWII

The Bretton Woods global financial institutions, which were created after WWII, fostered a rapid expansion of international markets. They succeeded beyond anybody's expectations. International trade expanded during this period three times faster than the world economy as a whole. This is what globalization is all about. Industrialization is resource intensive and it was fueled by inexpensive resources exported from developing nations, threatening their forests, minerals, and biodiversity.

These resources were and continue to be exported at very low prices. As a result, poverty grew in resource-exporting regions and provided "competitive advantage" in the form of cheap labor and cheap resources that exacerbated and amplified overconsumption in the North. Resources were over-extracted in poor nations desperate for export revenues, and over-consumed in industrial nations. Globalization after WWII increased together with an increasing global divide between the rich and the poor nations, the North and the South.[6] This is how the global financial system that was created by the Bretton Woods Institutions in

1945 is connected with the global environmental crisis we currently face. And this is how the global financial institutions caused an enormous global wealth divide between the North and the South.[7]

Energy is at the center of this process because energy use goes hand-in-hand with economic progress, and most of the energy used in the world today is fossil-based (87%). GDP growth is closely tied with carbon emissions today. Industrial nations with about 18% of the world's population consume about 70% of the world's energy. The North–South divide is therefore inexorably connected to the carbon emissions that are destroying the stability of our global climate. The same North–South divide has been a stumbling block in every United Nations negotiation on climate issues, for example in the 2009 Copenhagen Convention of the Parties (COP15) of the United Nations Framework Convention on Climate Change (UNFCCC), and then in 2010 in Cancun, Mexico COP16. This same issue surfaced in the Paris COP21 in December 2015. The problem is: who should use the world's resources: the rich or the poor? Or, otherwise put, who should abate carbon emissions?[8]

It can be said that we are reliving last century's Cold War conflict, but this time as a conflict between China and the U.S.[9] Each party could destroy the world, as they are the largest emitters and each alone can change the world's climate. Each wants the other to reduce carbon emissions (to "disarm") first. But this time the conflict is between the rich nations represented by the U.S. and the poor nations represented by China. The solution requires that we overcome the North–South divide, and that we restructure the use and trade of the world's resources between the rich and the poor nations. Otherwise put, global justice and the environment are two sides of the same coin. Poverty is caused by inexpensive natural resources in a world where developing nations are the main sellers of resources in the international market, resources which are over-consumed by the rich nations. This perverse economic dynamic is destroying the stability of the atmosphere, undermining climate patterns and causing the sixth largest extinction in the history of the planet.

How long will it take until this situation reaches its logical limits and victimizes our own species? How can we avoid extinction?

The Gordian knot that must be severed is the link between *natural resources, fossil energy, and economic progress*. Only clean energy can achieve this. But according to the IEA, this would require changing a US$ 45 trillion power plant infrastructure, the power plants that produce electrical power around the world, because 87% of world's energy is driven by fossil fuels and power plants produce 45% of the global carbon emissions.

Is it possible to make a swift transition to renewable energy?

Who Needs a Carbon Market?

Energy is the mother of all markets. Everything is made with energy: our food, our homes and our cars, the toothpaste and the roads we use, the clothes we wear, the heating of our houses and offices, our medicines: everything. Changing the cost of energy, making fossil fuel energy more expensive and undesirable and clean energy more profitable and desirable, changes everything. It makes the transition to clean energy possible. We have the technologies, we just have to get the prices right. Is it possible to thus change the price of energy?

Yes, it is. In fact, it has already been done, although it requires more input at present to continue and complete this process.

Here is the background and a summary of the current situation. In 1997, the carbon market of the United Nations Kyoto Protocol was noted and approved by 160 nations. In it, and after a long period of lobbying and designing the carbon market, one of us wrote the structure of the carbon market.[10] The Kyoto Protocol became international law in 2005, when the protocol was ratified by nations representing 55% of the world's emissions. The Protocol and its carbon market were adopted as international law. The U.S. signed the Protocol but it did not ratify it. The carbon market helped change the value of all goods and services in the world economy because it changed the cost of energy: made clean energy more profitable and desirable, and dirty energy unprofitable. This changed the prices of all products and services in the world, since everything is made with energy, driving the economy to use cleaner rather than dirty energy sources. It became more profitable and less costly to use clean energy that reduces emissions of carbon; this is precisely the role of the carbon market in the Protocol signed in Kyoto, in December 1997.

The carbon market started trading carbon credits at the EU Emissions Trading Scheme (EU ETS) in 2005 when it became international law. The World Bank reported on its progress in its report, "Status and Trends of the Carbon Market," which was published annually until recently. The carbon market requires carbon emissions limits to continue working. However, the 2015 Paris Agreement contains no carbon emissions limits — none at all — which is what is needed to avert catastrophic climate change. The World Bank documents that by 2011 the EU ETS was trading about $175 billion a year, and succeeded in decreasing the equivalent of over 20% of the EU's emissions of carbon. Through the carbon market, those nations who over-emit compensate those who under-emit, and throughout the entire process the world's emissions remains always under a fixed total emissions limit. These limits are for Annex I nations, and they are documented nation by nation in the Appendix to the Kyoto Protocol. Annex I nations are essentially OECD nations. A "carbon price" emerges from trading "carbon credits," or rights to emit, which represent the monetary value of the damage caused by each ton of CO_2. The carbon market therefore introduces a "carbon price" that corrects the negative impact that the emissions of CO_2 have on climate, which has been called "the biggest externality in the history of humankind."[11]

The carbon market cuts the Gordian knot and makes change possible. It does so because it makes clean energy more profitable and fossil energy less profitable, and therefore encourages economic growth without environmental destruction: it fosters green development. The carbon market itself costs nothing to run, and requires no subsidies except for minimal logistical costs. In net terms, the world economy is exactly in the same position before and after the carbon market: there are no additional costs from running the carbon market, nor are there any from its extremely important global services. Nations that over-emit are worse off, since they have to pay. But every payment they make goes to an under-emitter, so some nations pay and some receive. There are no costs to the world economy from introducing a carbon market, nor from the limits on carbon emissions and environmental improvement that it produces. It is all gain.

What is the status of the carbon market today? As of 2010, it has been ratified in 195 nations, and this includes essentially all the industrial nations except the U.S. It has been international law since 2005. Its nation-by-nation carbon limits expired originally in 2012 and were extended to 2015 and in a second period to 2020. But the Protocol itself — its overall structure and the structure of the carbon market — will not expire. They are and will continue to be international law. All we have to do to keep the carbon market's benefits is to define new emissions limits nation by nation for the OECD nations — something that we should be doing in any case, as they are major emitters, and without limiting their emissions there is no solution to global climate change.

What is the current status of the carbon market in the U.S., which is the single industrial nation that has not yet ratified the Protocol? There are cross currents in the U.S., since it is a politically divided nation. But the U.S. already has a carbon market for 10 northeastern U.S. States, called the Regional Greenhouse Gas Initiative (RGGI), which is operating, albeit timidly: the limits on emissions are small and thus so are the prices for carbon credits. The economic incentives of the Protocol's carbon market are enormous. China, for example, reportedly created one million new jobs and became the world's main exporter of clean technology, wind energy and solar equipment, after signing on and ratifying the Protocol; it also benefitted by about $75 billion from its carbon market and Clean Development Mechanism (CDM). China also introduced its own national carbon markets, although national or regional markets do not have the same status or positive effect in controlling climate change as the global carbon market does, as they are not based on global emissions reductions. Reducing global emissions of CO_2 is required in order to avert catastrophic climate change.

Many in the U.S. want part of the U.N. carbon market's advantages. President Obama said he wished to ratify the Protocol, and by now 22 states plan to create carbon markets of their own, including California, which has its own carbon market in operation. Hundreds of cities and towns support the carbon market in the U.S. In the Fall 2007, the U.S. Supreme Court agreed that the federal government and the Environmental Protection Agency (EPA) could enforce carbon emission limits without requiring Congressional approval. Every effort to deem this regulation illegal by Republican representatives has failed so far. It is generally accepted that global businesses (for example, the automobile industry) could benefit from Kyoto's guidelines, and could suffer economic losses without the benefit of Protocol's economic incentives at home. This is because the automobile industry is global, and cars that do not sell in other OECD nations create huge losses that can lead to bankruptcies. Since most OECD nations are buying carbon-efficient cars, because they ratified the Kyoto, the U.S. car industry could become commercially isolated. In part for these reasons, in 2010 the EPA imposed automobile emissions limits of 36.7 miles per gallon, an efficiency requirement that was increased further by the Obama administration in 2011 and again since then. The automobile industry voluntarily supported a rise to 54 MPG in 2011. And in December 2011, the EPA announced that it would impose limits on stationery sources like power plants, which is the first step towards a U.S. carbon market. The breakthrough Clean Power Act (CPA), a law created by President Obama and the EPA in 2014–15, imposed 30% reductions on power plants. However, several states contested this law and in 2016, in an unprecedented move, the U.S. Supreme Court froze its implementation pending the states' decisions. The issue is still hotly contested by the Republican Party, which stalled decision making since the 2016 election. Still, we must not forget that one of the most recent Republican candidates for president, Mitt Romney who was formerly Governor of Massachusetts, once endorsed the creation of a "cap and trade" system or a carbon market. A similar sequence of events took place when the SO_2 market was created at the Chicago Board of Trade (CBOT) over 20 years ago: first it was quite controversial, but SO_2 emissions limits were eventually imposed on U.S. power plants and then traded efficiently in an SO_2 market at the CBOT. This is widely considered to have been successful in eradicating acid rain in U.S.

Are the new EPA carbon limits the beginning of the U.S. carbon market as SO_2 limits were for the SO_2 market 20 years ago? Will President Trump succeed in eliminating the EPA and its environmental policy? History is being written right now.

Green Markets are the Answer — They will Transform Capitalism in the 21st Century

What is a green market and why does it matter? A shining example of a green market was just discussed: it is the Kyoto Protocol carbon market, which became international law in 2005. By 2011 the EU ETS was trading $175 billion annually and transferred about $130 billion in total to developing nations for clean technology projects that promote sustainable development. Most importantly it succeeded in its mission, EU emissions have decreased more than 20% since it became law in 2005. This happened while all other nations outside the Kyoto Protocol, such as the U.S., increased their emissions. Another successful example of a green market is the SO_2 Market in CBOT that was created about 20 years ago. This market is quite different from the carbon market because SO_2 concentration is not a "global commons," since it varies city by city while CO_2 is the same uniformly all over the planet. This changes fundamentally the structure and functioning of the market. There are other green markets in the works. The U.N. is exploring market mechanisms for biodiversity and for watersheds. As in the case of the Kyoto carbon market, these are markets that would trade rights to use the global commons — the world's atmosphere, its bodies of water, its biodiversity — and therefore have a deep built-in link between efficiency and equity. In the carbon market of the Protocol, by design, poor nations are preferentially treated, having in practice more access and more user rights to the global commons (in this case the planet's atmosphere). This is not the case with SO_2, which is a simple "cap and trade" approach as SO_2 is not a public good, as mentioned earlier.

Efficiency with equity is what green markets are all about. They are really two sides of the same coin: One is equity and the other is efficiency. Both matter. The carbon market provides efficiency with equity. How? Through its CDM the Kyoto Protocol provides a link between rich and poor nations, indeed the only such link within the Kyoto Protocol, since poor nations do not have emissions limits under the Kyoto Protocol and therefore cannot trade in the carbon market. Nevertheless developing nations have strong incentives for emissions reductions through the CDM of the carbon market — how does this work?

The CDM was described in a previous chapter and it works as follows. Private clean technology projects in the soil of a developing nation — for example in China, Brazil, or India — that are proven to decrease the emissions of carbon from this nation below a "U.N. agreed baseline" are awarded "carbon credits" for the amount of carbon that is reduced. These "carbon credits" are themselves tradable for cash in the carbon market, in recognition of the amount of carbon avoided in those projects. The carbon credits are a monetary compensation for clean technologies, and therefore shift prices in favor of clean technologies as the carbon market does. By law, CDM carbon credits can be traded for cash within the carbon market. This is the role of the carbon market in the CDM. This is how the CDM has provided about $130 billion in funding to developing nations since 2005.[12]

The North–South conflict, namely, who should abate first, puts all this at risk. To move forward in the global climate negotiations we must overcome the China–U.S. impasse, which is an intense form of the same conflict that prevails between rich nations and poor nations as a whole, the conflict between the North and the South.[13]

Is it possible to overcome the North–South divide? Yes, it is. But the interests of the industrial and developing nations are so opposed that once again, we need a two-sided coin. This is the same dual role that the carbon market played in the UNFCCC in 1997, allowing it to save the negotiations from which the Kyoto Protocol was born. The carbon market was acceptable to the rich nations because it provided market efficiency that the U.S. and the OECD wanted; at the same time the carbon market placed mandatory emissions limits solely on Annex I (OECD) nations' emissions, which is what poor nations wanted. This was what Chichilnisky saw then: how, by introducing the carbon market into the wording of the Protocol, it was possible to save the negotiations. This is how the Protocol was voted on by 160 nations in December 1997. Equity and efficiency are two sides of the same coin. Together they win. We need both.

The G20 and the rest of the world seem to recognize the need for sustainable development, in terms of both financial practices and the environment. In a nutshell Sustainable Development means giving the future a fair treatment in our policies. The concept of *basic needs* created in the Bariloche Model in 1974[14] is its backbone, and sustainable development is defined as the present satisfying its basic needs without depriving the future from satisfying its own needs.[15] The formal theory of Sustainable Development was created in 1993.[16]

We now turn to the principles and the practice of a new economic system that can achieve what is needed in the context of the global environment to avoid extinction.

Blueprint for Sustainable Development

In its creation, the G-20 stated as its top priority the achievement of Sustainable Development for the world economy. This requires:

(1) Economic growth in developing and rich nations to satisfy the *basic needs* of the present and the future;
(2) An accelerating transition to renewable energy and to a harmonious use of the earth's resources;
(3) Clean and abundant energy available worldwide.

Nobody knows the economic systems that will prevail in a long-term future. In the immediate future, however, sustainable development can be achieved by Green Capitalism, and we turn to discuss what this means in practice.

Green Capitalism

Green capitalism is a new economic system that values the natural resources on which human survival depends. It fosters a harmonious relationship with our planet, its resources and the many species it harbors. It is a new type of market economics that addresses both equity and efficiency.[17] Using carbon negative technology it is possible to reduce carbon in the atmosphere while fostering economic development in rich and developing nations, for example in the U.S., the EU, China and India. How does this work?

In a nutshell Green Capitalism requires the creation of global limits or property rights nation by nation for the use of the atmosphere, the bodies of water and the planet's biodiversity, and the creation of new markets to trade these rights. From this, new economic values emerge, as does a new concept of economic progress that goes beyond GDP, it is now understood to be needed.[18]

Green Capitalism can help avert climate change and achieve the goals of the 2015 U.N. Paris Agreement, which are very ambitious and almost universally supported but have no way to be realized within the Agreement itself. Green Capitalism is needed to achieve the requirements of the UNFCCC IPCC Fifth Assessment Report which focuses in averting catastrophic climate change. The Kyoto Protocol carbon market and its CDM play critical roles in the foundation of Green Capitalism; by using carbon negative technologies new economic values emerge to redefine GDP and remain within the world's "CO_2 budget."

Below are the building blocks for Green Capitalism and practical examples of how these organizing principles can be put in to practice right now. They illustrate how new carbon negative technologies can help achieve the climate negotiation goals, averting climate change.

Building Blocks for Green Capitalism

The three building blocks for Green Capitalism include:

(1) Global limits, imposed nation by nation, in the use of the planet's atmosphere, water bodies, and biodiversity,
(2) New types of markets to trade these limits, based on equity and efficiency, which are relatives of the Kyoto carbon market and the SO_2 market. Their prices create new measures of economic values and update the concept of GDP,

(3) Efficient use of carbon negative technologies to remain within the world's carbon budget and avert catastrophic climate risks providing a transition to clean energy and ensuring economic prosperity in rich and poor nations.

The building blocks have practical implications. On the basis of these blocks it is possible to resolve key goals of global policy. For example, we could create a $200 billion/year Green Power Fund from existing funding sources, including the Kyoto CDM, to ensure a smooth and accelerated transition to clean energy, achieving the goals of the U.N. Kyoto Protocol, the Paris Agreement and of the U.N. Green Climate Fund.

The building blocks offer practical ways to avert climate change and assist the ambitious goals of the Paris Agreement, which cannot be achieved within the Agreement's terms itself.

Indeed, according to the 2014 IPCC Fifth Assessment Report,[19] in practical terms carbon negative technologies, also known as carbon removal, are now needed on a massive scale in our century in order to avert catastrophic climate change.

Following are practical examples of how these building blocks can help achieve the goals of the UNFCCC, using carbon negative technologies while fostering growth in developing nations and overcoming poverty, all of which requires more energy supplies:

1. *Carbon negative power plants for developing nations*

New generation technologies can capture CO_2 from the atmosphere at low cost (http://www.globalthermostat.com).

These technologies can be used to build carbon negative power plants that clean the atmosphere of CO_2 while producing electricity.[20] Global Thermostat LLC is an award-winning firm that can be used as an example. The firm is commercializing a technology that takes CO_2 out of air and uses low cost residual heat rather than electricity to drive the capture process, making the entire process of capturing CO_2 from the atmosphere inexpensive. There is enough residual heat in a coal power plant that it can be used to capture twice as much CO_2 as the plant emits, thus transforming the power plant into a "carbon sink." For example, a 400 MW coal plant that emits 1 million tons of CO_2 per year can become a carbon sink absorbing a net amount of 1 million tons of CO_2 instead.[21] Carbon capture from air can be done anywhere and at any time, and so inexpensively that the CO_2 can be sold for industrial or commercial uses such as food and beverages, water desalination, greenhouses, bio-fertilizers, building materials and even enhanced oil recovery, all examples of large global markets and profitable opportunities. Carbon capture is powered mostly by low (85°C) residual heat that is inexpensive, and any source will do. In particular, renewable (solar) technology can power the process of carbon capture. It can help advance solar technology and make it more cost-efficient. This means more energy, more jobs, and also more economic growth in developing nations, all while cleaning the CO_2 in the atmosphere.

Carbon negative technologies can transform the world economy. In recognition of this fact, Global Thermostat has received several prominent awards.[22]

2. *The role of the Kyoto Protocol carbon market*

The role of the Kyoto Protocol carbon market and its Clean Development Mechanism (CDM) is critical, as it can provide needed funding and financial incentives for investment to build carbon negative power plants as described above, in developing nations. To provide access to all nations to the carbon market, the Kyoto carbon limits must be generalized to all nations, since no carbon market can operate without carbon emissions limits. The last chapter of this book explains how this can be achieved using carbon negative technologies. The CDM can be used to provide "off-takes," namely contracts that promise to buy the electricity that is provided by carbon negative power plants for a number of years. Using these off-takes as validation of future revenue unlocks banking resources for the investment required to build carbon negative power plants. Carbon negative power plants can be profitable, since their costs are low and they produce electricity. The scheme covers fixed costs and amplifies profits from clean technologies. Private capital markets recognize the business potential of clean technologies, having achieved a scope of US$260 billion a year in today's markets.

3. *The Green Power Fund and global capital markets*

To accelerate and enhance the impact of the U.N. carbon market and its CDM, it is possible to create a $200 billion a year private-pubic fund, also called the Green Power Fund. The fund can be used to build carbon negative power plants in developing nations, particularly in Latin America and Africa, enhancing their economic development while cleaning the planet's atmosphere. The Green Power Fund was named and proposed by Chichilnisky to the U.S. Department of State in Copenhagen COP15 in December 2009, and was published by Chichilnisky in the *Financial Times* in 2009. It was accepted by the U.S. State Department, and, two days later, then Secretary of State Hillary Clinton offered it up as the United States' contribution in the global negotiations in COP15. One part of it, called the Green Climate Fund — one word was changed — became international law and received substantial financial support. Most of the financial promises to the Green Power Fund, unfortunately, have not been delivered. The Green Climate Fund lacks the funding which Kyoto and its carbon market could provide if a link was made between the two. But the U.S. has not ratified the Kyoto Protocol and therefore has severed this natural and desirable source of funding. The connection can still be worked out while reinstating nation-by-nation carbon limits after 2020, as needed for the carbon market. The complete scheme, as proposed by Chichilnisky in COP15, is a private–public Green Power Fund. Funding can be raised from global capital markets to invest in investment grade firms that build carbon negative power plants in developing nations, with access to CDM funding to provide off-takes to buy the ensuing electricity.

The background and financial feasibility of the Green Power Fund can be seen as follows. Existing technologies[23] can efficiently and profitably transform coal power plants and solar thermal sources of energy into "carbon sinks" that reduce atmospheric carbon concentration while producing electricity. The more electricity is produced, the more residual heat is released, which drives the new generation carbon capture technologies.

The Green Power Fund provides the project finance that is needed to build carbon negative power plants in developing nations and elsewhere. This can accelerate the renovation of the \$45 to \$55 trillion power plant infrastructure worldwide (according to the IEA), which is 87% fossil today, to transform it into a powerful "carbon sink" that cleans the atmosphere of CO_2. Financially what is required is about \$200 billion/year for 15 years. By 2011, the U.N. Carbon

Market was already trading $175 billion/year, which almost suffices to cover these costs. The funding would go to investment-grade power plant builders[24] to build carbon negative power plants in developing nations. This is what the carbon market can trade per year, thus providing the funding required.[25] Therefore, the financial target proposed here is eminently achievable.

Green Capitalism and Traffic Lights for Human Survival

The building blocks proposed include new types of markets that can transform capitalism into Green Capitalism. This is achieved by transforming the economic values and prices of the new economy and providing market incentives that make green economic projects more profitable than their alternatives, fostering the conservation of biodiversity, clean water, and a safe atmosphere. Some of these new markets already exist and are described above. Green markets change GDP by valuing the global commons (the atmosphere, biodiversity, clean water), which in turn changes the measure of economic progress defined as the sum of all goods and services produced by an economy at market prices. As pointed out by *The Economist*,[26] well-known economists James Tobin and Bill Nordhaus gave examples showing that at present the measure of GDP "treats the plunder of the planet as something that adds to income, rather than a cost."[27] For example, cutting down all the trees in the U.S. national parks and making toilet paper from their wood, increases U.S. GDP and counts as economic progress. This is because GDP uses current market prices in its computations. Toilet paper has a market price, since there is a market for toilet paper, while there is no market for standing trees in the streets or in national parks.

How green markets change the measure of economic progress and redefine GDP

The creation of green markets that trade the use of the global commons, such as the rights to emit CO_2, to use drinkable water and to biodiversity, changes the measure of economic progress. The carbon market for example changes the GDP of a nation. For example, if two nations, which we can call Solar Nation and Coal Nation, produce exactly the same goods and services at the same cost but with the first using solar energy and the second coal, then the GDP of Solar Nation will be significantly higher than the GDP of Coal Nation on any given year. This is because if Coal Nation emits too much CO_2 it has to pay Solar Nation, which emits none. The difference makes Solar Nation's GDP higher and Coal Nation's GDP lower. In reality, the purchase and sale of carbon credits now enters the computation of GDP, giving a positive edge to Solar Nation and a negative one to Coal Nation. This is exactly what we wish to achieve: GDP that measures the damages that Coal Nation is doing unto the environment, the nation, and indeed the entire world.

Green markets that trade global public goods link equity with efficiency. As was explained, this is different from standard markets for private goods in which equity and efficiency are unrelated.

Examples of global green markets are:

- The U.N. carbon market (EU ETS), which has been international law since 2005.
- The SO_2 Market in the U.S., which started trading at the CBOT (Chicago Board of Trade) in 1991.

- Markets for Water and Markets for Biodiversity: these are still to emerge as proposed by Dr. Chichilnisky, and they are under U.N. consideration.

These markets provide missing signals that are normally provided by market prices when a good or service becomes scarce. Such signals are tantamount to traffic lights for human survival. Here are signposts to implement the above strategies going forward within the UNFCCC Global Climate Negotiations and the annual COP meetings, the last two of which were COP22 in Marrakesh in 2016 and COP23 in Bonn in 2017. We were able to insert the carbon market in December 1997 into the COP3 in Kyoto; in Copenhagen in 2009 we inserted wording into the CDM allowing carbon negative technologies to be compensated as part of the CDM, so that the CDM may fund negative carbon technologies, and four articles were inserted on carbon removals or carbon negative technologies in the Paris Agreement in COP21, December 2015. These are discussed in the next chapter.

Economic Incentives for the Short- and the Long-Run: Why Negative Carbon?

Long-run strategies do not always work for the short run as different policies and different economic incentives are required for the short run.

In the long run, the best policy is to replace fossil fuel sources of energy, which, by themselves, cause 45% of global emissions, with clean energy, and to plant trees to restore natural sinks of CO_2. However, fossil fuel power plants make up about 87% of global power plant infrastructure, worth about \$55 trillion. This makes it clear that fossil power plants cannot be replaced quickly, certainly not in the two or three decades in which we need to take action to avert catastrophic climate change. Once emitted, CO_2 remains in the atmosphere for hundreds of years, and we have emitted so much that unless we actually remove the CO_2 that is already there, we cannot remain long within the carbon budget, the concentration of CO_2 beyond which we expect catastrophic climate change.[28] In the short run, therefore, the IPCC indicates in its 2014 Fifth Assessment Report that we must actually remove the carbon that is already in the atmosphere and in massive quantities, this century.[29] This is a carbon negative approach that works for the short run. Renewable energy is a long run solution.

Renewable energy is too slow for the short run, since replacing a $45 trillion power plant infrastructure with renewable plants could take decades. We have already seen that planting trees is not feasible either, for similar reasons. We need action sooner than that. For the short run, we need carbon negative technologies that capture more carbon than what is emitted. Trees do that — and they must be conserved to help preserve biodiversity. Biochar does that, but trees and other natural sinks are too slow for what we need today, as previously discussed.

Negative carbon leads to a global economic transformation

Negative carbon is needed *now* as part of a blueprint for transformation. It is needed for Sustainable Development and its short term manifestation, Green Capitalism. In the long run, only renewable sources of energy will do, including solar, wind, biofuels, nuclear, geothermal, and hydroelectric energy. Of these, only solar energy suffices to replace fossil fuels; the rest are in limited supply and cannot replace fossil fuels. Global energy today is roughly divided as follows: 87% is fossil, namely natural gas, coal, and oil; 10% is nuclear, geothermal, and hydroelectric, and less than 1% is solar power — photovoltaic and solar thermal. Nuclear fuel is scarce and nuclear technology is generally considered dangerous as per Japan's tragic experience in 2011, at the Fukushima Daishi nuclear power site. It seems unrealistic to seek a solution in the nuclear direction. Only solar energy can be a long term solution at present: less than 1% of the solar energy we receive on Earth can be transformed with existing technologies into 10 times the fossil fuel energy used in the world today.

Yet we need a short-run strategy that accelerates long run renewable energy, or we will defeat long-run goals. In the short run, as the IPCC validates, we need carbon negative technology. The short run is the next 20 or 30 years. There is no time in this period to transform the entire fossil infrastructure — it costs $45 trillion (according to the IEA) to replace. We need to directly reduce carbon in the atmosphere now. We cannot use traditional methods to remove CO_2 from smokestacks (often called Carbon Capture and Sequestration, CSS) because they are not carbon negative as is required. CSS works but does not suffice because it only captures what power plants emit currently. Any level of emissions adds to the stable and high concentration we have today. We need to remove the CO_2 that is already in the atmosphere. What is needed is direct air capture of CO_2 which the IPCC calls carbon removals.

One solution is to combine air capture of CO_2 with storage of CO_2 into biochar, cement, polymers, and carbon fibers that can replace a number of other construction materials such as metals.[30] It is also possible to combine CO_2 to produce renewable gasoline, namely gasoline produced from air and water. CO_2 can be separated from air and hydrogen separated from water, and their combination is a well-known industrial process to produce gasoline. There are also new technologies using algae that make synthetic fuel commercially feasible at competitive prices.

Other policies involve combining air capture with solar thermal electricity using the residual solar thermal heat to drive the carbon capture process. This could make solar plants more productive and efficient as a source of energy.

In summary, the blueprint offered here is a private-public approach, based on new industrial technology and financial markets. It is self-funded and could use carbon credits and the Kyoto Protocol CDM, fostering mutually beneficial cooperation between industrial and developing nations. The blueprint provides two sides of the coin, equity and efficiency.

Our vision is a carbon negative economy with green capitalism resolving the global climate crisis and the North–South divide. In the examples provided above, carbon negative power plants and the capture of CO_2 from air together ensure a clean atmosphere coupled with economic development: as more energy and jobs are created, the cleaner the atmosphere becomes.

In practice, Green Capitalism means economic growth that is harmonious with the Earth's resources.

A Vision for Sustainable Development

Avoiding extinction means protecting the survival of the human species. Survival is not about violent competition and struggle. It is about life not death. Carbon negative solutions are the future of energy, and green markets lead the way to Green Capitalism, resolving the global climate crisis and the global divide, while providing clean energy and economic growth for the North and the South that is harmonious with the Earth's resources. It means creating and nurturing life and building a sustainable future.

9

Four Obscure Articles in the Paris Agreement Hold the Key to Resolve Climate Change

After months, if not years of preparation, and two weeks of intensive negotiations, the Paris COP21 climate conference in late 2015 produced the Paris Agreement[1] hailed by world leaders as "a turning point for the world," signifying the end of the fossil fuel era. With no time lost, much of the news industry responded in a similar fashion, describing the agreement as the first legally binding global climate deal[2] representing "the best chance to save the planet."[3] Unfortunately, these hopeful statements are not supported by the facts. The first legally binding international agreement in climate change was the Kyoto Protocol when it became international law in 2005. In contrast, the Paris Agreement has essentially no mandatory elements: it requires only the disclosure of intended nationally determined contributions (INDC) which are voluntary and do not suffice to meet even half of the targets of the Agreement. Furthermore, the Paris Agreement requires no action for at least five years. Overall, the Paris Agreement scales up hopes and ambitions by including nearly all nations; it is a statement about the seriousness of climate change and of hope about solutions, but it has no action plan. With the recent threat of U.S. withdrawal, it is not a universal agreement either.

On Earth Day, April 22, 2016, around 60 heads of state gathered at the United Nations in New York. About 175 governments took the first step of signing the Paris Agreement that day, and at least 34 nations representing 49% of the greenhouse gas emissions formally ratified it. By comparison, the Kyoto Protocol was ratified by nations representing over 55% of global emissions in 2005 when it became international law.[4]

The April 22, 2016 signing has been interpreted as a declaration that we have reached the end of the age of fossil fuels. In his address to the assembly at the United Nations headquarters in New York, U.N. Secretary-General Ban Ki-moon argued, "We are in a race against time. The era of consumption without consequences is over."[5]

Ultimately, however, the Paris Agreement falls short. With no action plan and no mandatory provisions, we may be no closer to averting catastrophic climate change than we were before the start of the Paris talks. In this sense, this could be one of the worst failures of the global climate negotiations in their 24 years of existence.[6]

Will this be the verdict of history? The story does not end there. There is something new and unexpected.

The Paris Agreement incorporates four articles which have gone largely unnoticed but which can offer a lightning rod to solve the climate change crisis. These articles contain the seeds of global policies that can cut through the maze of political, economic, and technological issues in which the world seems to be mired. The four new articles indicate what is needed now to execute the Paris Agreement and resolve the climate crisis. For example, Article 4.1 of the agreement states: "In order to achieve the long-term temperature goal… Parties aim to reach global peaking of greenhouse gas emissions as soon as possible…and to undertake rapid reductions thereafter in accordance with the best available science, so as to achieve a balance between anthropogenic emissions by sources and removals by sinks of greenhouse gases in the second half of this century…" What is new here is the introduction of carbon "removals" to reach a solution.

Explaining in Some Detail to Help Make it Happen

Chichilnisky participated officially within a party delegation of the Paris COP21 conference and in that role helped introduce into the Agreement the four articles that incorporate *carbon removals* from the atmosphere. These articles are somewhat technical, but have powerful economic and political ramifications that could turn the world economy on its head. And this is the moment when we need to do precisely that. Let's not underestimate the importance of what is at stake. "Carbon removals" refers to removing the carbon that is already in the atmosphere. This matters because the Intergovernmental Panel on Climate Change (IPCC) Fifth Assessment Report has found that carbon removals are needed in this century on a massive scale to avert catastrophic climate change. In most scenarios, without carbon removals there is no way to avert climate change. It's as simple as that. We know that we procrastinated too long in taking action, and that CO_2 remains in the atmosphere for hundreds of years once emitted. For those reasons, it does not suffice anymore to decrease CO_2 emissions into the atmosphere. We have to remove what is already there. Indeed, as the latest IPCC Fifth Assessment Report points out,[7] carbon emissions reductions are not enough to avert global warming. According to the Report, in most scenarios we are headed with certainty towards an increase in temperatures by 3°C by 2100, and many scientists believe that 2°C of warming is already "a recipe for disaster." It suffices to recall that the North and South Poles are melting. Closer to home is Superstorm Sandy, which closed down New York City for weeks and flooded its streets and subways, leaving entire neighborhoods without electricity, including schools and law enforcement stations, and stranding automobiles floating in the streets. Climate change means an increase in the frequency and severity of such climate events. This means, for example, three or four Superstorm Sandys every year in New York; even this proud city could not survive such climate change. On May 3, 2016, the front page headline of the *New York Times* read "Resettling the First American Climate Refugees," an article about a $48 million grant for Isle de Jean Charles in Louisiana, the first allocation of federal tax dollars to move an entire community struggling with the effects of climate change.

The conclusion from the 2014 IPCC Fifth Assessment Report is that in addition to drastically reducing emissions and adopting clean energy systems, it is now imperative to remove existing carbon dioxide from the atmosphere. And now a major international agreement, the Paris Agreement, contains not one but four articles referring to the necessary carbon removals that can transform everything.

A Fortuitous Coincidence Can Change Everything

The following is is an account by Dr. Chichilnisky, one of this book's authors:

The story goes back to 1996 when I was U.S. Lead Author for the IPCC. It continues in 1997 when I introduced the wording for the carbon market into the Kyoto Protocol, and in 2009, when I introduced the Green Power Fund in Copenhagen COP15, which was accepted by then Secretary of State, Hilary Clinton, to become the Green Climate Fund and international law. In 2006, I had discovered the need for carbon negative technologies to avert climate change. To tackle this, I co-invented a technology now operating in Silicon Valley that takes CO_2 directly from the air. Because of this work, I was invited to participate in the Paris COP21 meetings of December 2015 as an official adviser to the Papua New Guinea COP Delegation as well as to a 50-nation group called the Rainforest Coalition, which comprises almost 25% of the U.N. vote. In that role, and with great help from Kevin Conrad and Federica Bettio, both official representatives of the Panama delegation to COP21 and leaders of the Rainforest Coalition, we were able to include "carbon removals" or "carbon negative technology" as part of the official "mitigation actions" in Articles 4.1, 4.13 and 4.14 of the Paris Agreement. Both rich and poor nations are now obliged to report carbon removals as part of their mitigation actions in the COP21 Intended Nationally Determined Contributions (INDCs) going forward. In addition, developed nations had pledged to "provide," "take the lead," and "scale up" finance for developing nation's mitigation in Articles 9.1 and 9.4 of the Paris Agreement, and this was defined to include carbon negative technology or carbon removals. Carbon negative technologies such as those employed by Global Thermostat in Silicon Valley, California are well positioned to help, since removals must be reported and can be funded under the provisions for mitigation.[8] The next step could be to determine if any further "guidance" could be required under the COP. If so, an agenda item can be opened, and the Subsidiary Body for Scientific and Technological Advice (SBSTA) committee of the UNFCCC could meet and develop such guidance, if one is needed. The existing Green Climate Fund could then start financing carbon removals. The Paris Agreement mentioned, without defining its sources, that $100 billion in financing is needed to compensate poorer countries for "loss and damage," mitigation, and adaptation. Much more is needed, of course, to ensure that poor nations don't pay the price for decades of guzzling, coal burning, and emissions by the richest countries. This is just a start.

Why does this all matter? How can we use the leverage that these articles provide? One needs to understand the entire picture first.

The delegates attending the Paris COP21 conference, to their credit, acknowledged the problem of unchecked climate change: *this is the main achievement in Paris, a nearly universal acknowledgment of the problem.* But can a climate catastrophe be avoided by voluntary and universal acknowledgment, by expression of hope among nations to reduce emissions and keep temperatures from rising further? The pact agreed to in Paris on December 12, 2015 by nearly 195 nations is voluntary, it has no teeth. The agreement is bound to nothing and no action can be taken until 2020. Yet the four articles that we helped to introduce can make a difference to the climate crisis. They contain the seeds of the entire solution, and can precipitate a global transformation in the world economy needed for resolving climate change.

The Paris Agreement holds "the increase in the global temperature average to be well below 2°C above pre-industrial levels" and has been called a "dangerous equivocation." Recall that the only mandatory provision in the Paris Agreement is for nations to provide reports on the progress of their Intended Nationally Determined Contributions (INDCs), whatever those may be, and only starting in 2020. And even if the INDCs were mandatory — which they are not — all together they do not suffice to meet even half of the voluntary targets. In contrast to the Kyoto Protocol, the Paris Agreement includes no legally binding carbon dioxide emissions limits. As stated, there aren't any mandatory emission limits nor mandatory payments in the Paris Agreement to help poor nations develop clean energy technologies, or to mitigate the damages already caused by climate change, or to repair the damage historically perpetuated by rich nations. But the four new articles (4.13, 4.14, 9.1 and 9.4) can turn things around. To see how, we need to show how to go from failure to success. The worst failure of the Paris Agreement is the lack of mandatory emissions limits on the signatory nations, which are necessary for the carbon market to operate. What is traded in the carbon market is the right to exceed one's mandatory limits. With no mandatory limits, there can be no carbon market. The entire world is clamoring for a "price on carbon": this is provided by the carbon market. Even the largest oil and gas companies in the world publicly supported a price on carbon (including Shell, BP, Statoil, Total and Engie). Yet the Paris Agreement undermines the very foundation of a price on carbon by requiring no mandatory emissions limits.

How can four new articles help recover mandatory emissions limits that are needed for the successful functioning of the carbon market? Here is the answer: with carbon negative technologies, also called carbon removals, both the rich and the poor nations can and will accept *conditional* mandatory emissions limits, from which mandatory limits will naturally evolve. How this happens is shown below. With mandatory emissions, the carbon market can function and make funding available to implement carbon removals. To make all this possible the world needs carbon negative technologies that can inexpensively remove CO_2 from air and transform the CO_2 into goods and services that trap it on the planet itself. Below, we show that there are carbon negative technologies that can remove carbon and sell it for building materials, polymers, food and beverages, greenhouses, synthetic fuels, and agents to desalinate water and produce safe bio-fertilizers — all of which are economically viable and can make a profit. Yes, we can capture CO_2 in the atmosphere and transform it into profitable goods and services, while cleaning the atmosphere. The global CO_2 markets are, or will soon be, large enough to absorb all the CO_2 that humans emit into the atmosphere today, which is over 32 gigatons/year. New technology enables carbon removals, helps the economy, creates jobs, and improves exports. We can even create carbon negative power plants that turn the energy sector into a carbon sink, cleaning the atmosphere of CO_2 and enabling the Paris goals just mentioned: keeping temperature increases below 2°C. The circle closes. Carbon removals enable economic growth and economic growth enables carbon removals. That is how carbon removals can help achieve the goals of the Paris Agreement and resolve climate change.

In reality, as the IPCC validates, carbon removals are the only way we can now avert catastrophic climate change: they are therefore a necessary and sufficient condition to overcome climate change.

Then what are we waiting for? Why has this not happened yet? The answer is simple: the technology is new and we are talking about a revolution in the use of energy, which turns power production and economic growth into a way to clean the atmosphere. It seems impossible and too good to be true. But it is true.[9]

The next step is to explain *conditional mandatory limits* and how they facilitate the continuation of mandatory emission limits by exceeding them to all nations. Conditional limits are emissions limits that poor nations will accept because they are conditional on receiving funding to deploy them. In the case of rich nations, these conditional limits are conditional on the availability of profitable technologies that help the economy create jobs and exports while removing CO_2. Nobody has been against such conditional limits, and they can catch like wildfire once people see them with their own eyes and implement them.

Make no mistake: this is an economic revolution. It is now possible to ensure that the more energy we use, the more we clean the atmosphere and avert climate change. All of this is simple and obvious, and so far I have found no resistance to the idea, even though the innovation is so profound that its consequences take one's breath away. The company, Global Thermostat, for example, is commercializing its technology and has already received seven awards as one of the world's most innovative companies in energy. Once in place, conditional mandatory limits lock themselves into mandatory limits, thus breathing life into the existing U.N. carbon market and continuing the role of the Kyoto Protocol beyond 2020. This, in turn, makes funding readily available from the Protocol's CDM and the carbon market itself to implement carbon negative technologies, or carbon removals, therefore meeting the "conditions" we started from. The circle closes. The carbon market was already trading \$175 billion/year by 2011 so the funding of \$200 billion/year that the Green Power Fund requires can be readily available with conditional emissions limits and a well-functioning carbon market. For the developing nations, this means CDM funds, of which the carbon market has already provided about \$200 billion. For all of this to happen, we need carbon removals and a carbon market, and for this, in turn, we need carbon negative technology and mandatory limits. We have the technology, so this locks together into place as a gigantic puzzle: the global economy being first unscrewed and then screwed up again, but in the opposite direction. Too good to be true? This is the political, economic, and financial structure that is needed. It overcomes poverty and creates sustainable economic growth, saving humankind from a climate catastrophe that can cause our extinction. Not too bad. All of this is coming from four new and obscure articles of the Paris Agreement.

A key factor in making this real is a technology that can implement carbon negative solutions, namely carbon removals, in a commercially viable way. There are now carbon negative technologies in Silicon Valley, like those employed by Global Thermostat, that are operating at SRI, Menlo Park California. Global Thermostat's plant at SRI cleans the CO_2 from SRI power plants powered by natural gas, and removes additional CO_2 from the air.[10] These technologies can offer a solution to the greatest threat facing human civilization as we know it. They are called disruptive technologies for obvious reasons.

In terms of global policy, to implement this in a time-scale that matters, we must activate the EU ETS carbon market based on mandatory emissions and already trading $175 billion/year in 2011. The funding from the carbon market would suffice to quickly scale up carbon removals around the world, as the IPCC now requires. As an example, this can occur through carbon negative power plants that clean the atmosphere while they produce electricity — such as that in SRI — and do this at a low cost and in a profitable fashion. A proposal Chichilnisky made in Copenhagen COP15 in 2009 was to use the Kyoto carbon market to offer finance to scale up carbon negative power plants in poor nations, thus providing the electricity that is needed by 1.3 billion people worldwide who lack access to it. This process would clean the planet's atmosphere at the same time. This was called the Green Power Fund, and it required $200 billion/year for building carbon negative power plants; instead the Green Climate Fund was created in COP15 and later made into international law, changing one word (Power to Climate) but severing its connection to its main source of funding, the carbon market of the Kyoto Protocol.

Why did the Green Climate Fund sever the connection to its ideal source of funding, the carbon market of the Kyoto Protocol, which could still offer enough funding to resolve climate change? This happened for the same reason that the Paris negotiations ended in an agreement with no teeth. The reason is the long standing insistence by U.S. Congress — through its venerable and unanimously voted for Byrd-Hagel Act of 1996 — on no mandatory emissions limits. However, now, for the first time, using carbon negative technologies, Byrd-Hagel conditions for emissions limits can be met. How?

The technology now exists to remove carbon from the atmosphere, as required by the IPCC, while making profits, creating jobs, and enhancing exports. The technology needed is already operating in Silicon Valley. We also have a financial model — the Green Power Fund — that can use the funding from the carbon market of the Kyoto Protocol (about $200 billion/year) to resolve the climate change crisis. All we need is to implement, politically, the financial and technological solution within our power.

The whole thing hinges on mandatory emissions limits. We even have a political solution to implement mandatory limits in a way that complies with the requirements of the venerable Byrd-Hagel Act, in the form of conditional mandatory limits. For developing nations, this solution provides funding on the condition that carbon removal projects are implemented, which makes the projects profitable and leads to the acceptance of mandatory limits. We can also have conditional mandatory limits for industrial nations, conditional on their being achieved through profitable carbon removal technologies. Everybody benefits, and we can save human civilization from disappearing under the rising oceans.

In summary, the 2015 Paris Agreement can be seen as a statement of hope by the nations of the world, without an action plan to implement what it advocates. However, it is possible to solve the problem using four articles in the Paris Agreement that support "carbon negative technologies," such as those implemented by Global Thermostat. This, and similar technologies, can build "carbon negative power plants" that are profitable, providing energy for development and poverty alleviation in poor nations, while cleaning the atmosphere, as required by the IPCC to avert catastrophic climate change. The U.N. carbon market suffices to fund this effort, and will do so once we extend the conditional emissions limits to mandatory emission limits: the carbon market can then accelerate the removal of carbon from the atmosphere, as is needed to avert climate change.

The Green Climate Fund is all we have so far, and it has rather limited funding, as it has no identified source of revenues. So far, the most reliable source of funding for clean technology — and the largest scope ever of such funding — has been the UNFCCC carbon market (EU ETS) and its CDM. It is therefore important that the carbon market be allowed to continue its excellent work, since carbon removals are critical to prevent catastrophic climate change according to the 2014 IPCC Fifth Assessment Report.

The carbon market can only operate if there are legal limits on carbon emissions in the nations that trade in the carbon market. Carbon negative technology — namely "carbon removals" — can help achieve this goal. Through conditional emissions limits, the carbon market can fund carbon removals, and carbon removals can help implement the carbon market. This is a self-implementing positive cycle that augurs well for the global climate. All developing nations leaders I know have expressed their interest and would accept conditional emissions limits. They can be compensated to implement the limits with the newest technologies. For rich nations, emissions limits can be contingent on the implementation being economically feasible, costing less than the price obtained by selling the CO_2 removed from the atmosphere, and even making profits. The combination of conditional carbon limits for rich and poor nations and carbon negative technologies would suffice for the carbon market to work. And the CDM, which draws funds from the carbon market, can offer the $200 billion/year that is needed to start funding the Green Power Fund, as defined above. The result will be to implement in a timely fashion carbon removals, which we know are needed to avert catastrophic climate change.

We can see a future for humankind, with a stable climate. In a new global economy, the CO_2 removed from the atmosphere can be an enormous source of wealth. It can service the production of building materials, carbon fibers, polymers, and water desalination agents; it can augment food production in greenhouses and provide CO_2 for carbonated beverages; and it can help to grow the economy while cleaning the atmosphere. It is a new world of economic possibilities, opening before our eyes, where CO_2 replaces petroleum as the basic feedstock of the 21st century. Limits on emissions, water consumption, and the use of biodiversity, can create new market values. This can be sustainable development in the most authentic sense of the word. The carbon market and its relatives, global markets for water and for biodiversity, can create prices that value the global commons and change economic values forever. With the new prices, we can align economic progress with human survival. The dream of a future with appropriate values is within our reach. Let's do it.

10

Reversing Climate Change

The pace of global climate change is accelerating. So is our understanding of what it takes to avert catastrophic climate change. The world's power plant infrastructure is the largest source of emissions, which is set to increase dramatically: power plants globally represent about US$55 trillion in global revenue today, and by 2050, energy use is widely expected to double in developing nations.

In the long run, we must produce energy from clean sources that do not emit CO_2. That is clear. However, the implementation of this long-run solution will take decades, and we are running out of time. Once emitted, CO_2 can remain for hundreds of years in the atmosphere. It does not decay fast. The United Nations is urging significant action on CO_2 emissions in the next 12 years. We don't have time to wait. Other solutions are needed right away.

October 2018: A Seismic Shift

The handwriting is on the wall. Previous chapters explained the solutions and summarized many years of policy and technology work. But no matter how clear and compelling the solutions may be, the innovation required is radical and will take a long time to be absorbed and understood, and even longer to be adopted and executed.

After years of relentless work facing incredulous reactions, it all converged at one point in time: October 2018. Three public scientific reports published by the U.N. Intergovernmental Panel on Climate Change (IPCC), the U.S. National Academy of Sciences (NAS), and the U.S. Global Change Research Program (USGCRP), announced the many years of efforts previously considered unrealistic had been unexpectedly validated. The results were explained in an accessible article that appeared in *The New York Times* on October 24, 2018. These four publications,[1] together, explained and documented that removing CO_2 from the atmosphere is now the only solution for averting catastrophic climate change, and that the technology exists now to reverse climate change. These startling announcements agree with the earlier findings presented in this book just prior to publication. With time running out to avoid dangerous global warming, the U.S.'s leading scientific body — the U.S. National Academy of Sciences — set out to explore options for solution. On October 17, 2018, it urged the federal government to begin a research program focused on developing technologies that could remove vast quantities of carbon dioxide from the atmosphere in order to help decelerate climate change. A 369-page report was written by a panel of the U.S. National Academies of Sciences, Engineering and Medicine.[2] It underscored an important shift. For decades, experts had said that nations could prevent large temperature increases mainly by reducing reliance on fossil fuels and moving to cleaner sources like solar, wind hydraulic energy and nuclear power. Even the Kyoto Protocol and its carbon market were designed to reduce or limit CO_2 emissions. But at this point, nations had delayed so long in cutting their carbon dioxide emissions that even a breakneck shift toward clean energy would most likely not be enough. Something else must be done now: that is, the physical removal of the CO_2 that is already in the atmosphere.

The second landmark scientific report appearing in October 2018 was issued by the leading global authority on climate change, the UN Intergovernmental Panel on Climate Change (IPCC), which was awarded the Nobel Prize for its work. At that time, Chichilnisky acted as Lead Author of the IPCC representing the U.S. The report concluded that taking out a big chunk of the carbon dioxide already loaded into the atmosphere may now be necessary to avoid significant further warming.

As indicated in previous chapters, there are technologies that can achieve this, although it is fair to say that there is no universal agreement on how to do so economically, or at sufficient scale. And the IPCC report made it clear that we will have to do it fast. To meet the climate goals laid out under the Paris Agreement, humanity may have to start removing around 10 billion tons or gigatons of carbon dioxide from the air each year by midcentury, in addition to reducing industrial emissions, said Stephen W. Pacala, a Princeton climate engineer who led the panel. That is nearly as much carbon as all the world's forests' and soils' current absorption each year. "Midcentury is not very far away," said Dr. Pacala. "To develop the technologies and scale up to 10 billion tons a year is a frightful endeavor, something that would really require a lot of activity. So the time would have to be now." The panel's members conceded that the Trump administration may not find the climate change argument all that compelling, since the president has disavowed the 2015 U.N. Paris Agreement. But it was quite likely that other countries would be interested in carbon removal. Nevertheless the United States could take a leading role in developing technologies that could one day be worth many billions of dollars. Despite President Trump's somewhat dated position, in February 2018 the U.S. passed a bipartisan federal law, named the "Future Act" (45Q), providing a $35 tax credit per ton of CO_2 removed from air, in practically unlimited amounts. Chichilnisky was requested to write the proposal for this law by Senators Sheldon Whitehouse (D-RI) and John Anthony Barrasso (R-WY) in late 2017, both of whom proceeded to draft, negotiate and pass the bill into a bipartisan federal law, becoming part of the federal budget. It became the 45Q federal law in February 2018. Aside from the carbon market of the Kyoto Protocol, which became international law in 2005, this U.S. federal law is the most advanced environmental legislation of its type, as it addresses directly the need to remove massive amounts of CO_2 from the air and offers attendant financial incentives. Most existing legislation is restricted to avoiding emissions, which no longer suffices to resolve climate change.

But how do we execute on this plan — how do we do what it takes? It has been difficult enough so far to reduce CO_2 emissions. How do we reverse the process entirely and remove the CO_2 that is already in the atmosphere as needed to reverse climate change?

Right now, there are plenty of ideas for carbon removal kicking around, and previous chapters have discussed some of them. Countries could plant more trees that pull carbon dioxide out of the air and lock it in their wood. Farmers could adopt techniques, such as no-till agriculture, that would keep more carbon trapped in the soil. The United Nations demonstrated however, that natural carbon "sinks" such as vegetation, and related policies are too slow and too few for resolving the problem and as has also been discussed in previous chapters. Often called BECCS, these technologies were found to be unable to remove the amount of CO_2 needed. In any case, as the National Academies panel warned, many of these methods are still unproven or face serious limitations. There is only so much land available to plant new trees. Scientists are still unsure how much carbon can realistically be stored in agricultural soils. The most recent U.N. study found that there isn't enough land to do the job, and that the use of land for this purpose could compete with the land needed for agriculture, to feed the world's human population which is expected to reach 9 billion by 2050. In theory, it might be possible to collect wood or other plant matter that has absorbed carbon dioxide from the air, burn it in biomass power plants for energy and then capture the carbon released from combustion and bury it deep underground, creating, in essence, a power plant that has negative emissions. While no such facilities are operating commercially today for this wood and plant matter solution, the technology to build them exists. A serious potential problem with this approach, as we discussed earlier in the book and which was raised by the National Academies panel, is that the land required to grow biomass for these power plants could run into conflicts with the need for farmland for food. The panel estimated that this method might one day be able to remove 3 to 5 billion tons of carbon dioxide from the air each year, but it could possibly be much less, depending on land constraints. That is a far cry from the 10 to 20 billion tons of carbon dioxide we may need to pull out of the air each year by the end of the century in order to limit overall global warming to around 1.5°C (2.7°F), according to the recent United Nations report. That figure assumes nations manage to decarbonize their energy and industrial systems almost entirely by 2050. If nations fail to hold global warming below that 1.5°C level, the United Nations report warned, tens of millions more people could be exposed to life-threatening heat waves and water shortages, and life on the planet will be seriously harmed — as an immediate example, the world's coral reefs could disappear almost entirely.

In view of this, as documented above, a few companies are building "direct air capture" (DAC) plants that use chemical agents to scrub trace amounts of carbon dioxide from the air, allowing them to sell the gas to industrial customers or bury it underground. Most direct air capture plants are still too expensive for mass deployment. Global Thermostat has the only technology acknowledged by the U.S. National Academy report in its technological description — which agrees with Global Thermostat's patents — to eventually and sufficiently lower the cost per ton of CO_2 from the air down to $18 per ton. This would put the process in the realm of physical reality and, equally important, in the realm of commercial reality that would greatly facilitate its execution.

The National Academy's panel recommended a dual strategy. The United States could set up programs to start testing and deploying carbon removal methods that look ready to go, such as negative emissions biomass plants, new forest management techniques or carbon farming programs. At the same time, federal agencies would need to fund research into early-stage carbon removal techniques, to explore whether they may one day be ready for widespread use.

As the *New York Times* article reports, for instance, scientists have long known that certain minerals, like peridotite, can bind with carbon dioxide in the air and essentially convert the gas into solid rock. Researchers in Oman have been exploring the potential to use the country's vast mineral deposits for carbon removal, but there are still major questions about whether this can be done feasibly on a large scale. In its report, the panel laid out a detailed research agenda that could ultimately cost billions of dollars. But given that carbon removal could "solve a substantial fraction of the climate problem," the report said, those costs are modest. For comparison, the federal government spent $22 billion on renewable energy research between 1978 and 2013. Until now, all federal funding went to the removal of CO_2 from industrial chimneys or "flues," which contain a 5% to 7% concentration of CO_2 but this does not remove CO_2 in net terms as is now required. The old fashioned form of CO_2 capture cannot do the job since it is, at best, "carbon neutral," leaving the atmosphere as is now — while what is now needed is "carbon negative technology"™ that leaves the atmosphere with a strictly lower concentration of CO_2. The new technology that removes CO_2 from air is very different: in air, the concentration of CO_2 is 400 parts per million which is 3,000 times more dilute than in industrial chimneys' flues. No federal funds have so far been provided for achieving this important and necessary step to clean the atmosphere. Global Thermostat has raised $52 million in investment since September 2010, already completing three CO_2 plants that produce CO_2 from air.

Outside experts hailed the U.S. NAS report as a sign that carbon removal is finally becoming central to discussions about tackling climate change. "We're moving from the early stage of 'what is carbon removal?' to figuring out what specific steps can be taken to get these solutions at scale," said Noah Deich, executive director of the group Carbon180, which recently began an effort to bring researchers and companies together to advance carbon removal technologies in the marketplace. The National Academy's panel did, however, warn of one potential drawback of carbon removal research. It could create a "moral hazard," in which governments may feel less urgency to cut their own emissions if they think that giant carbon-scrubbing machines will soon save the day. To that end, the panel stressed that carbon removal, if developed, could only be a part of a larger global warming strategy. "Reducing emissions," the report noted, "is vital to addressing the climate problem." The need for reducing CO_2 emissions is something that previous chapters of the book have underscored as well.

A Game Changer: Capturing CO_2 from the Air

In the wake of the three October 2018 reports about the need to remove CO_2 and for Direct Air Capture, the authors decided to add this brief chapter to the book to summarize the status of climate change solutions, and in particular Global Thermostat and its breakthrough Direct Air Capture technology.

Founded in 2010 by Graciela Chichilnisky and Peter Eisenberger, Global Thermostat (GT) was created with the mission of reversing climate change while making profits. The reasoning is that time is of the essence to reverse climate change, and a solution that is commercially viable is likely to be implemented faster. Furthermore, such an economic model is sustainable by itself and can be implemented without government subsidies. In a world where government strategies are somewhat challenged, this matters a great deal. As co-founder, Chichilnisky, saw GT as a way to help implement the carbon market she had written into the U.N. Kyoto Protocol in 1997, as the removal of CO_2 could facilitate achieving the mandatory national carbon emissions limits that are at the main objective of the Kyoto Protocol and of its carbon market. The profitable removal of carbon could also facilitate the provision of energy needed by developing nations as is explained in the last section of this chapter on carbon negative power plants. CO_2 can also be used to produce desalinated water biofertilizers and even clean synthetic gasoline by mixing CO_2 with hydrogen. GT technology therefore facilitates development based on the satisfaction of *basic needs* in developing nations, a concept that Chichilnisky introduced and which was voted on by 150 nations as the main concept of economic development in the Earth Summit of the United Nations, Rio de Janeiro, 1992. *Basic needs* is also the cornerstone of the concept of sustainable development for which she created the formal theory. They'll benefit from the fact that CO_2 is available in the same concentration all over the atmosphere, inducing a form of equity in the availability of a key input for energy production and economic development. Eisenberger, a world renowned material physicist who researched at the famous Bell Laboratories, led R&D globally at Exxon and then served as Director of the Lamont Doherty Earth Observatory, a professor of physics at Princeton University, a Vice Provost of Columbia University and the founding director of the Earth Institute at Columbia University. Global Thermostat's founders met their lead investor Edgar Bronfman Jr. and told him their idea, which they patented, to create and commercialize a technology that could clean the atmosphere while making profits, whereupon Edgar Bronfman said: "This is too good to be true, and usually things that are too good to be true do not work out." Nine years later, with critical support from him and several other investors, it has become clear that it is true. As Forbes and KPMG said, GT technology could reverse climate change; it can clean the planet's atmosphere while making profits, a genuine economic revolution that can help avert catastrophic climate change. Eisenberger and Chichilnisky have agreed that doing so will not cost the world economy, but it involves a switch to a Renewable Energy and Materials Economy in a way that can bring forth a period of unprecedented global prosperity, energy security and climate stability. Since its creation in 2010 and in pursuit of this mission, Global Thermostat has raised \$52 million and created, developed and patented a proven breakthrough technology that can cost-effectively remove CO_2 directly from the atmosphere. It is, therefore, called a carbon negative technology™. GT's multi-patented "Direct Air Capture" (DAC) technology is owned individually by the firm's co-founders, its co-inventors, and is licensed to Global Thermostat on an exclusive basis for commercial exploitation. GT's technology is now commercially available and has the potential to fundamentally change the way we manage and utilize carbon on a global scale, including the use of CO_2 in instead of petroleum, which represents an economic revolution of sorts. Since CO_2 emissions are considered a leading cause of climate change and current concentrations in the atmosphere (400 ppm) are at their highest level in the last 3 million years according to the NOAA and Scripps, GT has the power to turn this major liability into a valuable commodity that can serve an addressable CO_2 market, spanning a wide range of industries, including water desalination, biofertilizers, food and beverages, production of polymers and carbon fibers, synthetic fuels, enhanced oil recovery, and more — a rapidly growing CO_2 market that McKinsey & Co., the global consultancy, anticipates can generate $1.1trillion annually.

The principal advantages and characteristics of the Global Thermostat technology are as follows: (1) it provides CO_2 that can be monetized as a replacement for fossil fuels; (2) it has low capital costs; (3) it has a scalable design that is highly modular; (4) it has a low variable cost that represents a cost breakthrough, as it is powered by free or low-cost, widely available residual low temperature (85°C) heat; (5) it slashes transportation costs, as CO_2 can be produced where used, requiring no transportation; (6) it is a carbon negative solution that suffices to redress climate change, since the captured CO_2 can be sequestered in carbon fiber, into "aggregate" and other building materials that can replace steel, aluminum, and concrete and costs less than the materials replaced, a market that can absorb 10 gigaton of CO_2 per year; (7) it has easy integration into upstream energy production or downstream CO_2 utilization.

For Global Thermostat the key to achieving commercial value is its low cost for capturing CO_2 from air. GT's technology was characterized in technical terms by the U.S. NAS October 2018 report as allowing, with sufficient scale, an eventual cost of about $18 per ton of CO_2 removed from air. GT can capture CO_2 where it can be transformed into

GT's second Commercial Demo plant

valuable products without requiring pipelines or other expensive forms of transportation to reach their markets, at a lower cost than existing oil or gas-based alternatives. As already explained, GT can provide practically unlimited low-cost clean CO_2 produced anywhere, and without transportation costs. In addition to public reports, such as those of the U.S. NAS, several private top-tier third-party technology reports have validated that GT's carbon capture technology is currently the lowest cost production method for CO_2 due to its ability to utilize low temperature residual process heat (at 85°C) as its main source of energy, its proprietary sorbents and its unique system of supports for the sorbents. By 2019, GT has been awarded over 60 patents and, by filing under the international Patent Cooperation Treaty, it has patent protection in 152 countries. The potential of GT technology to capture CO_2 from air at an economically viable cost led the company to partner with several well-known industrial partners clients and investors, including classic beverages companies, top-tier oil producers and engineering companies worldwide. As part of its mission, GT's top priority is the rapid and deep expansion of plants that capture CO_2, and to offer preferential pricing policies and deep discounts for developing nations.[3]

How does Global Thermostat Technology Work?

Here is a snapshot of how GT's technology works: Air and/or the flue gas mixture are moved by fans over a wall of honeycomb monoliths, which are coated with a proprietary 'sorbent' (amine-based chemical). The coated monoliths adsorb the CO_2. Steam produced from residual heat sources is used to desorb the CO_2 from the wall. High purity CO_2 is recovered. A Concentrated Solar Power (CSP) power plant can be used to drive the process, making it most effective in terms of carbon removal while producing CO_2 free electricity.

The purity level of the CO_2 gas recovered by GT's process is as high as 99%, and the CO_2 stream exits at one atmospheric pressure; it can be further purified and/or liquefied using standard "compression" techniques, to be used for food and beverages applications.

Since transportation costs for large volume gaseous CO_2 are significant and can run as high as $1.5 million per mile for a pipeline, plus compression, producing CO_2 on site drastically reduces or eliminates these costs. A CO_2 air capture plant can be located anywhere, needing only air and heat to operate. It can be built next to an oil field or a food processor, and it can be located anywhere there is air, thus eliminating the need to truck or pipe the CO_2 over a long distance.

The basic operational steps of the GT capture system are analogous to those of a giant dehumidifier and are summarized here, described in more detail below:

Step 1: Air Input
Ambient air is driven through the "monolith contactors" to adsorb CO_2; these are similar to the monolithic convertor parts that are currently used in most automobiles on earth.

Step 2: Carbon Capture
GT's proprietary sorbent has a high affinity for CO_2, it is a polymer impregnated within the monoliths that captures the CO_2.

Step 3: Desorption and Regeneration
The CO_2-rich sorbent is heated with low-temperature process heat steam (85–105°C) releasing the CO_2.

Step 4: Heat Management

The clean, emptied sorbent and cooled down monolith is now ready to repeat the process, capturing more CO_2.

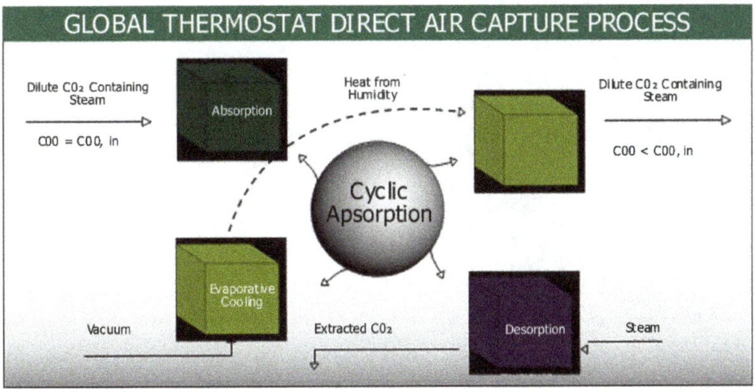

The GT capture system: Global thermostat direct air capture process

GT technology: technical information and commercial performance

In high-level terms, the technology consists of a unique, proprietary, low-cost process to capture CO_2 from the atmosphere (at concentrations of 400 parts per million). It can also capture CO_2 from industrial flues or chimneys (with 5%–7% concentration or more), or from any combination of the two. As of 2019, GT has been awarded 61 patents in the US and internationally and by treaty it is protected in 152 nations. These patents cover the entire process for the low-cost loading and unloading of CO_2 from a porous block monolith coated with proprietary sorbents to produce >98.5%–99% pure CO_2 gas stream at atmospheric pressure.

The unique GT technology and application of the GT capture process uses advanced techniques and chemistry and ensures process differentiation vis-a-vis conventional systems — a fact borne out by GT's patents, and a Freedom to Operate legal opinion. The process is highly effective and

The Carbon Capture Unit

The photograph on the left is of one specific embodiment: the Carbureted Embodiment from the Commercial Demonstration plant at SRI, in Menlo Park, CA, in operation connected underground to SRI's power plant, capturing CO_2 from March 1, 2013 to the present.

The principal advantages and characteristics of the GT system as a whole are:

- *Low Capital Costs*
- *Scalable design:* modular design can be adapted for different sized applications; plants can ramp up from 2,000 tonnes/yr. up to 3mm tonnes/yr. or more, simply by adding more modules.
- *Low Variable Cost Breakthrough:* powered by free or low-cost, widely available residual process heat; slashes transportation costs as plant can be located near the point of demand.
- *Carbon Negative Solution:* GT plants capture CO_2 from emissions as well as the atmosphere, thereby enabling the capture of more CO_2 than the point sources emit.
- *Flexible integration:* can be retrofitted to legacy power plants, new power plants, industrial facilities (e.g. cement, steel), regardless of whether they are

A second Commercial Demo plant, connected underground to SRI power plant is up and running, spring 2013. Approximate dimensions are 10 ft. W × 12 ft. d × 35 ft. H.

efficient, even at low atmospheric concentrations of CO_2 and at ambient temperature and pressure. This has been verified by a GT Pilot Plant operating at SRI since October 2010; a second, larger and more advanced GT plant at SRI in Silicon Valley has been operating since March 2013 as well as in several bench-scale demonstrations — all documented in third-party laboratory reports by Georgia Institute of Technology, Carmagen, BASF, SRI and, most recently, by a 2018 U.S. National Academy of Sciences report.

Technology operation: each plant is a series of modules

The basic operational steps of the GT capture system are illustrated here:

Step 1: Air Input
Ambient air and/or flue gas is driven through the monolith contactors to adsorb CO_2.

Step 2: Carbon Capture

GT's proprietary sorbent (polymer chemically bonded to the monoliths) captures the CO_2. This is effective even though CO_2 is very dilute in the air (400 ppm).

"Contactors" (the inner surface of the monoliths) provide high surface contact areas at low pressure drop, enabling movement of large volumes of air with effective contact of CO_2 without needing to apply too much force, and therefore keeping operating cost down.

Step 3: Desorption and Regeneration

The CO_2-rich sorbent is heated with low-temperature process heat steam (85°–95°C).

The steam serves multiple roles: heat source, pressure front, and sweep gas. It strips the CO_2 from the monoliths, where it is collected and piped away for use (at nearly standard pressure & temperature).

The clean, emptied sorbent is now ready to repeat the process, capturing more CO_2.

To improve performance, the timing of the adsorption and regeneration cycles are optimized and matched.

Step 4: Heat Management

The clean, emptied sorbent and cooled down monolith is ready to repeat the process, capturing more CO_2. Steam condensate is flash evaporated off the hot regenerated monoliths and condenses on an adjacent bed that just finished Step 2 to preheat it and save energy.

The system employs a counter-cycling tandem module configuration — such that while one stack (a set of 2,000+ monoliths) is adsorbing CO_2 the other is desorbing and regenerating — thus ensuring a steady and continuous

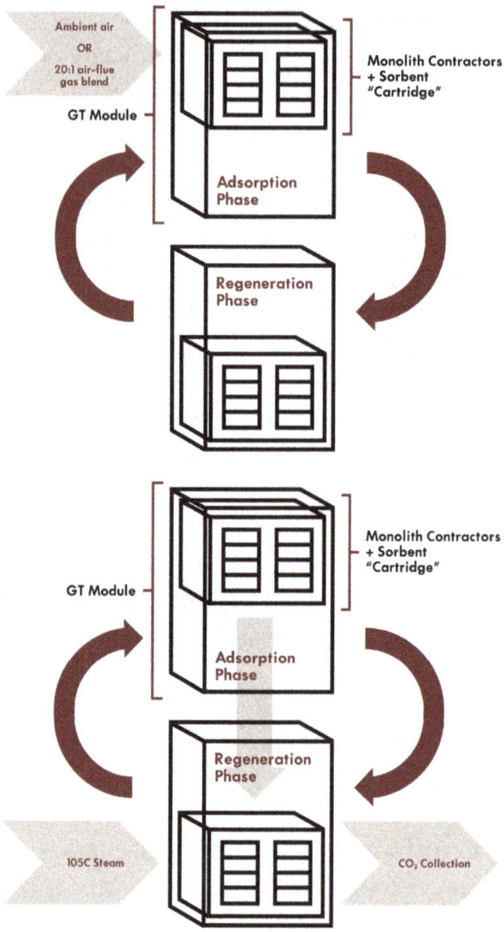

capture process. In this configuration, there is continuous adsorption and maximum carbon dioxide capture.

Three technology embodiments

GT developed the technology to be modular, scalable, and adaptable to the source of CO_2. Accordingly, GT has three primary embodiments of the technology:

1. ***Pure Air or Direct Air Capture Embodiment:*** this focuses on capturing CO_2 at atmospheric levels, without

access to an enriched source like an industrial chimney. Such a system can be built in any location and requires a source of power for example 85°C residual heat in order to operate.

2. ***Carbureted Embodiment:*** a GT plant captures from more concentrated carbon sources that can be 7–10% CO_2: it is installed at the site of existing CO_2 emissions, such as by attachment to the flue of legacy and new power plants, industrial facilities (e.g. glass, steel), and similar locations of high CO_2 concentration and waste heat.
3. ***Self-Carbureted Embodiment:*** an additional source of CO_2 and heat (i.e. a gas turbine) is purposefully installed coincidentally with a GT plant, which is built at or near the point of CO_2 consumption. This facilitates customized CO_2 systems that satisfy demand on site, slashing transportation costs. This allows for high quantities of CO_2 to be captured for remote users.

Containerized units

Industrial gases provide large, existing markets for CO_2, but there exist a wide array of smaller distributed markets for CO_2 with much higher profit margins. For these applications, the scale of CO_2 consumption for a single installation is much less than the 50,000 or 100,000 tonnes per year that GT's full-size commercial modules can deliver. These markets represent a low capital barrier of entry with high revenue returns. GT has developed smaller "containerized" modules to effectively penetrate these markets, so its units are effectively contained within standard shipping containers. These markets are potentially large in volume but highly distributed, and include — but are not limited to — applications like beverage carbonation, water desalination, greenhouse agriculture, specialty materials production, etc.

GT plants or units consist of containerized modules that are so called because they are constructed within standard ISO shipping containers. This allows GT to pre-fabricate the

Model of DAC module contained within a standard 40-foot ISO shipping container

complete unit in the factory, and then ship the container to its deployment site without any additional construction — simply by connecting heat, electricity, water and flue gas for the Carburetor embodiment, the module is ready to operate. This not only saves a significant amount of cost in the form of on-site labor but allows each and every containerized module to be built in the same location with the same quality standard. The automated production of factory-fabricated modules drives the cost down significantly.

The Containerized Carburetor embodiment consists of a single 40-foot ISO shipping container that is capable of producing 10,000–15,000 tonnes of CO_2 per year depending on location and conditions. The Containerized DAC embodiment consists of two 40-foot ISO shipping containers and a third 20-foot ISO shipping container for auxiliary equipment, capable of producing 2,000–5,000 tonnes of CO_2 per year. As with the full-scale embodiments, multiple modules can be operated simultaneously to achieve plant capacities as multiples of the base units — however, at large enough scale it can become more cost effective to build a plant consisting of full-scale GT modules instead.

Comparison with conventional carbon capture system

Conventionally, CO_2 in ambient air, which is now about 400 parts per million and is uniform throughout the planet's atmosphere, has been considered too dilute to be captured economically. As a result, the focus of CO_2 capture applications has been on flue gas and other point sources of CO_2 in which there is up to 300 times more CO_2 concentration than in the atmosphere. However, there are also significant disadvantages with this conventional approach, including:

CO_2 captured at high-concentration sites requires transportation to the point of use. However, there is limited pipeline infrastructure, and new pipelines cost on the order of \$1.5 million per mile.

- Capture of CO_2 solely from flue gas often requires a cleanup of impurities (such as NOx and SOx), which adds to capture costs.[4]
- Conventional monolith/sorbent combinations require high heat for release and thus expensive cooling to cycle.
- Flue gasses tend to be hot, but CO_2 is most efficiently captured at lower, ambient temperatures.

GT's ability to capture unlimited amounts of CO_2 from ambient air creates efficiencies by allowing more mass to pass through a larger surface area. GT also blends flue gas with ambient air, optimizing CO_2 concentration and temperature. Consequently, GT provides:

- Shorter cycle times
- Lower temperatures for efficient CO_2 absorption and desorption
- Cleaner input gas (namely air)

- Reduced transportation costs

In these ways, GT challenges the accepted wisdom about direct air capture and offers the following in comparison with conventional carbon capturing:

- Lower operating costs
- Lower capital costs
- Can be located practically anywhere
- Flexible co-location with low-temperature residual heat sources — avoiding the need for costly CO_2 piping or trucking
- Lower barriers to adoption for existing emitters, as GT's technology can easily be retrofitted into plant infrastructure and/or integrated with conventional carbon capture technology, increasing yield and lowering costs
- GT reduces and/or eliminates transportation costs, as the CO_2 can be produced where required

The following section is a brief review of the technology divided into two sections: The Capture Process and the Regeneration Process.

Capture process — performance metrics

The capture process can be divided into the monolith contactor and sorbent components. The critical parameters are the amount of CO_2 captured and the cost of materials, each measured per square meter.

Timing

Within the capture and regeneration process, the timing of the respective phases is crucial. Conceptually, the capture phase could either be longer than, shorter than, or equal to the timing of the regeneration phase. To optimize the duty cycle (so that there is continuous CO_2 capture and release, which is important for some applications), timing should be equivalent, and a tandem design is used. In this design there are matched monolith stacks moving in opposite directions, so there would always be one stack capturing and one releasing.

Monolith contactor design and performance criteria

The first step in designing the technology is to select a monolith contactor. The use of monoliths for CO_2 capture has been patented by GT. This is a device that allows the gas stream to move through its open channels; first contacting the porous wall surface of the contactor, and then diffusing into the pores of the walls, where an embedded sorbent immobilizes the CO_2. The performance of the contactor determines how effectively CO_2 is removed. In visual terms, think of the monolith as a six-inch ceramic cube, which looks solid when viewed from above or the side, but when viewed from the front looks like a cross-hatch lattice of open space — as if you were holding hundreds of drinking straws edge-on (where the straws' diameters have been reduced to just one millimeter). The internal walls of those open channels are highly porous and impregnated with the chemical sorbent. As air passes through the channels and over the sorbent-covered walls, the CO_2 is captured.

Capture efficiency and process costs are the critical metrics. Capture efficiency — a main driver of cost — is proportional to the surface area contacted by the gas stream.[5]

Effectiveness of the monolith contactors is a balancing act. A key goal is to ensure a minimum amount of pressure is needed to push air through the contactor channels (called "pressure drop"). This must be traded off against factors such as optimizing the exposure time of the sorbent to the air (ensuring carbon dioxide is optimally captured). This can be at odds with a fast-moving airflow facilitated by low-pressure requirements.

GT's initial monoliths were composed primarily of cordierite (a ceramic); however, GT's next generation monoliths utilize alumina, to take advantage of superior characteristics (in weight, thermal mass, and pressure drop) to improve performance. There are tradeoffs between these two materials in cost, sorbent deposition, etc. Working with Corning, GT has developed proprietary designs that optimize the monoliths for these factors. As a strategic supplier, Corning works exclusively with GT in the realm of carbon capture.

Because of the low concentration of CO_2 in ambient air, a very large volume of air needs to be moved through the monolith to conduct pure air capture. In places where there is persistent wind (5-plus miles/hr.), one would not need to artificially push the air through the contactor, reducing operating costs. Alternatively, the contactor can be designed to minimize pressure drop, at a cost of cell density and capture capacity.

One can think of the contactor surface providing friction to the gas stream, so the more surface the gas passes over the more resistance it experiences. The effective resistance can vary greatly depending on whether the airflow is turbulent or laminar (smooth and ordered). Modeling studies by Carmagen and others reveal that a gas stream moving over the walls in the laminar, low-resistance regime provides the most contacting surface per pressure drop. This dictated that the channels through which the air moves must be straight. This process is critical for low costs and has been patented by GT as well.

A second and more fundamental tradeoff is between the need to maximize the carbon captured per square meter versus the increased force required to move more air. This tradeoff can be viewed as being between the capital cost per square meter (which decreases with increasing velocity and capture efficiency), and the operating cost (which increases with the higher velocity of the input stream). GT's technology is highly tunable, so GT can optimize the opex/capex balance to meet the requirements of any specific implementation.

Another important performance criterion is the speed of CO_2 diffusion into the contractor's porous walls, so that all the sorbent sites are effectively accessed.

During the early design phase of the technology, a large variety of contactors were evaluated, aiming to maximize surface area per pressure drop and speed of CO_2 diffusion. A monolith contactor with rectangular channels was discovered to be clearly superior — presenting minimal frontal surface area and promoting laminar flow. When comparing actual test data against modeling projections, it became clear that the gas dynamics of the capture process were well understood and could be reliably modeled. The fact that the system is performing as expected enables us to use standard equations to optimize the performance of the monolith in terms of the basic properties of the contactor (cell density, void fraction, wall thickness, length, and loading of CO_2 per volume of wall). GT addressed all these variables, resulting in a high-performance, proprietary contactor design.

Sorbent performance criteria

As noted above, an important balancing criterion for the sorbent is that it provides a high CO_2 capture capacity (maximizing the effective area for CO_2 loading into the pores of the monolith walls), while maintaining a sufficiently fast diffusion rate during the regeneration/desorption phase.

Crucially, GT has learned that the CO_2 diffusion and capture processes do not happen at a uniform rate; they actually happen very quickly, when the regenerated monolith/sorbent is first (re)exposed to the gas stream, after which the reaction slows down, as the monolith becomes saturated with CO_2. This has allowed GT to reduce the cycle-time for the entire process by a factor of 2–3, allowing GT to complete more full cycles per unit time, thus capturing more CO_2 per unit time, or per unit of capex, significantly increasing capacity while reducing costs.

Additionally, the sorbent material must also be small enough to fit into the pores of the monolith, but large enough so that it does not volatilize (break down due to excessive heating) in the temperature range of the process. Finally, the sorbent must maintain its effectiveness over an extended period of time, and thousands of capture/regeneration cycles. Sorbent life-limiting forces include: oxidation; volatilization; destruction of the pore structure by improper steam exposure during regeneration; and deactivation due to NOx & SOx exposure in flue gases.

GT has conducted accelerated lifetime studies, which show no loss of activity over the study period. Further extended lifetime studies are currently underway. GT has also determined that its latest sorbent formulations are essentially immune to NOx, and their SOx exposure degradation is mild — happening early in their lifetimes and then dropping off, such that sorbent levels can be "overfilled" initially, so that when the SOx degradation stops there will still be an optimum amount of active sorbent remaining. GT screened various sorbents at SRI and Georgia Institute of Technology (GIT) and found that Primary Amines were the best performers at the low partial pressure of CO_2 in ambient air. PEI, an amine polymer produced by BASF and others, was found to be the most effective sorbent, even though it has only 40%

primary amines. It is also commercially available at low cost ($10–$20/kg.). Other, even more advanced amine formulations are under development by GT and our partners at Georgia Institute of Technology.

Regeneration process

The major objectives of the Regeneration process are to:

1. Liberate the sorbent-bound CO_2
2. Collect the CO_2 at high purity
3. Return the monolith to ambient temperature

The amount of CO_2 collected in each cycle is called the Working Capacity. The performance criteria are to:

1. Minimize energy use
2. Produce high purity CO_2 that can readily be transported or utilized on-site
3. Maximize working capacity

For a given sorbent formulation, the regeneration (CO_2 stripping) process is common to all versions of GT's technology. Though the concentration of CO_2 in ambient air can be 250 times lower than that in carbureted (flue gas) air, once captured, the result is a CO_2-loaded sorbent that must be converted to pure CO_2 gas.

An important metric for this process is the temperature differential between the absorption/capture and desorption/regeneration states: the lower the temperature of the steam needed in the regeneration process, the lower the cost (and the more sources of residual steam become economically available). Thus, GT's ability to utilize low temperature residual process heat (at 85°-95°C) is a key advantage and differentiator, protected by patents.

Once the CO_2 has been heat-stripped from the sorbent-impregnated monolith, the monolith must be returned to initial conditions in order to capture more CO_2. In addressing this final step of regeneration, there are four objectives that need to be addressed. The first is to minimize energy use. It is therefore desirable to collect and reuse as much of the heat as possible. The second objective is to minimize water use, by recovering as much of the water condensed in the monolith as possible. Third, one needs to cool the monolith below 70°C before exposing it to oxygen and beginning the capture phase again. Finally, one needs to accomplish that cooling quickly so that it does not add significantly to the cycle time, which is directly related to the cost (i.e. the faster the cycle time, the more CO_2 a given monolith can capture per year, driving down capital costs).

GT carried out cooling scoping experiments to determine the effectiveness of evaporative cooling of the monolith/sorbent, learning that, by passing air over the monoliths and evaporating the condensed water, they can be cooled in less than 10 seconds.

Another objective of regeneration is to collect the CO_2 at high purity. Critical to achieving the >98.5% + target purity is evacuating the air from the monoliths before collecting the CO_2 and doing so with minimal mixing with the CO_2- laden sorbent. GT has effectively addressed all these concerns with its proprietary, US-patented regeneration process.

GT technology — third party validation

The GT technology was successfully demonstrated at three levels: at the single monolith level at SRI Laboratories; at the first pilot plant at SRI from 2010 to 2011 using 660 monoliths arranged as a single monolith stack; and at the second commercial demonstration plant at SRI that went operational in March 2013 to test and validate the Carbureted Embodiment, using twin monolith stacks. It relies on low-cost heat and steam from SRI's own cogeneration plant. Depending on the materials used, this plant can capture up to 10,000 metric tonnes of CO_2 per year. The key features of this plant are:

- Off-the-shelf cordierite monolith with alumina coating and PEI sorbent;
- Demonstrated 50% CO_2 capture efficiency;
- Over 1,000 hours of total operation, including 200-plus hours continuous operation with no obvious oxygen deactivation or any other loss of effectiveness.

Laboratory tests conducted by the Georgia Institute of Technology in Atlanta, SRI in Menlo Park, California; the company BASF; and the U.S. National Academy of Sciences have verified a technology evolution roadmap for reducing capture costs by a factor of 2 or more.[6] This includes optimizing monolith composition and geometry, the primary amine sorbent, cycle timing, and balance of system materials and structures. The design optimization through tandem stack design and carburetor concept (for 20x CO_2 production via 97-to-3 ambient air-flue gas blend) drove the technology well down the cost curve. Developing a paired module system, whereby neighboring machines share steam dramatically reduced operating costs. On the Direct Air Capture (DAC) side, developing a system whereby a single regeneration chamber can be in constant utilization, by services multiple monoliths stacks, has also made the DAC system much more efficient and affordable.

Carbon Negative Power Plants[7]

It is known that power plants are the main source of CO_2 emissions, comprising about 45% of the world's emissions. What is less known is that the combination of conventional power plants with GT technology can transform the world's US$55 trillion power plant infrastructure into a sink of CO_2

emissions, cleaning the atmosphere. A conventional fossil fuel plant produces enough residual heat that it can drive a GT carbon capture plant to capture about twice as much CO_2 as the power plant emits itself in the first place. For example, a coal driven 400 MGW power plant emits between 1 and 2 million tonnes of CO_2 per year; the residual heat the power plant produces is generally enough however to capture twice as much CO_2 as the plant emits. The following are representative input figures that make the point:

Heat electricity water and land footprint per ton of CO_2

While inputs will vary somewhat for each Global Thermostat plant, the following are representative figures:

Electricity	260 kWh/MT
Heat	4.8 MMBTU/MT
Water	0.1 MY/MT
Land	1.09 sq ft – 1.5 sq ft /MT

A brief company background is as follows: Global Thermostat LLC (GT, www.global thermostat.com) was formed in 2010 to develop and commercialize a unique technology for the direct capture of carbon dioxide from the atmosphere and other sources. The GT process "co-generates" carbon capture with other industrial processes — such as power production — by using the residual heat from those processes to drive its carbon capture technology. By combining CO_2 capture from air along with capture from the flue gas of an electrical power plant, and using the power plant's low-cost residual heat to provide the energy needed for the air capture process, GT technology has the capability of transforming power plants into net carbon sinks. Global Thermostat technology also can work with renewable power plants, because it captures carbon directly from air using the plant's process heat. For example, heat from a Concentrated Solar Plant (CSP) can be used by Global Thermostat to drive its capture process. This also creates a second source of revenue for the power plant which can now sell CO_2 as well as electricity.

CO_2 capture from air, or direct air capture (DAC), is different from other forms of carbon capture in that it extracts CO_2 directly from the atmosphere at low temperatures and at a concentration of about 400 parts per million. Other carbon capture technologies typically extract CO_2 only from flue gases at higher temperatures and parts per million. Global Thermostat technology also can work very economically in conjunction with standard capture technology by combining air and the flue as its sources of carbon.

The U.S. Department of Energy recently announced \$2.3 billion in funding in conventional carbon capture technologies that capture CO_2 from concentrated CO_2 sources (e.g. from the flue of power plants, which is about 7–10% CO_2). Additionally, China has built into its Twelfth Five Year plan a 17% reduction of CO_2 per unit of GDP output and is funding a number of its own initiatives to accomplish this goal.

GT's technology is significantly different, as it can remove CO_2 from conventional sources as well as from direct air, which has about 400 parts per million of CO_2. This has multiple advantages. The most important are: (i) GT's plant location flexibility allows CO_2 capture where CO_2 can be used as a product, thereby reducing transportation and distribution costs, (ii) GT technology has the ability to make a power plant carbon negative, and (iii) GT technology uses low cost process heat to provide the energy needed for the air capture process.

CO_2 air capture has gained momentum on the policy front and in the business community as a viable and economic solution for reducing carbon emissions and is now being introduced commercially with pilot demonstration plants. As was already mentioned, the first GT pilot plant erected at SRI International in Menlo Park, California, captures 1,000 tons per year of CO_2 (although by changing its parts this can increase up to 10,000 tons per year and was co-developed with Linde, Corning and BASF. The CO_2 captured at plants like this is available for use in applications such as enhanced oil recovery, greenhouses, production of industrial grade formic acid, production of bio-fuels from algae, and, when combined with hydrogen, production of hydrocarbons such as high-octane gasoline.

According to the International Energy Agency, about 45% of all human based emissions of CO_2 are generated by power plants, and 89% of electricity production around the world is powered by fossil fuels. This represents an energy infrastructure valued in excess of $55 trillion U.S. dollars. As this cannot easily be replaced, CO_2 emissions will be around for some time to come. However, with CO_2 air capture, much of this emitted CO_2 can be recovered and molecularly tied up, thereby lowering the CO_2 load in the atmosphere. With GT's technology, the more electricity one produces, the more CO_2 one can reduce from the atmosphere. This reverses the paradigm that links fossil fuel-based power production with carbon emissions. A GT plant utilizing process heat can capture up to twice the CO_2 that a coal power plant emits, leading to carbon negative electrical power production.

Global Thermostat technology can reduce emitted CO_2 by 200% (it is carbon negative) and also can operate alongside other conventional methods of CO_2 capture. GT's technology enhances the efficiency, capture rates, and CO_2 purity levels generated by conventional processes, which typically only reduce 90% of the emitted CO_2. The Global Thermostat process holds significant value in reducing environmentally damaging CO_2, while creating economic value for CO_2 consuming industries such as beverages, biofertilizers, greenhouses, polymers, cement "aggregate", carbon fibers, synthetic fuels, and others.

Global Thermostat plant at SRI Menlo Park California produces about 1,000 tpy 99% pure CO_2 since 2013.

Epilogue: The Future Act of 2018: New U.S. Law Provides Unlimited Tax Credits to Remove CO_2 from Air

On November 2017, the new film "Carbon Negative" was successfully screened in New York City. The film was produced by Patrick Furlotti and John Stember, and directed by this book's Emmy Award-winning director Paul Atkins; in it one of this book's authors, Graciela Chichilnisky, explains the global risks we face and the policy implications of carbon negative technologies, including Global Thermostat's, with its unique ability to remove massive amounts of CO_2 from the atmosphere at low cost, effectively reversing global climate change as the U.N. IPCC says is now needed. The New York film screening was followed by a successful Q&A conducted by Chichilnisky. She was approached by Wil Gregersen, who works with Rhode Island Senator Jeanine Calkin (D-RI). Senator Calkin invited Chichilnisky to Rhode Island for a similar film screening followed also by a Q&A, to take place at the prestigious Rhode Island School of Design in Providence; Senator Calkin also arranged a brief presentation by herself and Chichilnisky at the Rhode Island Senate Chambers on how Global Thermostat and its unique technology could further Rhode Island's own sustainability goals. It turned out that Sustainability was a hot legislative issue in the U.S. state of Rhode Island.

Following the success of the two Rhode Island events, Chichilnisky was invited to Washington, D.C. to meet and discuss with U.S. Senator Sheldon Whitehouse (D-RI), who was working at the time on a carbon capture bill and was interested in hearing more about Direct Air Capture (DAC) of CO_2 and how it can be utilized commercially to further economic development, increasing jobs and exports while cleaning the atmosphere. After a detailed discussion Senator Whitehouse requested a proposal on what Global Thermostat could do for Rhode Island and for the U.S. as a whole, in terms of sustainability and increasing jobs and exports, and specifically how a Global Thermostat factory in Rhode Island could build, sell and ship Global Thermostat plants from Rhode Island to the entire world. Senator Whitehouse correctly surmised that given the urgent need for such solutions as those offered by Global Thermostat (unique in the U.S.), the market opportunities would eventually be enormous — trillions of U.S. dollars. Chichilnisky accepted the request, wrote a proposal for Senator Whitehouse and sent it to the Senator and his staff in DC in January 2018. Following this proposal, in February 2018 Direct Air Capture (DAC) became the subject of a new bill, the Future Act (45G), which became a bipartisan federal law (introduced mostly by Senator Whitehouse) defining for the first time DAC within U.S. legislation and offering unlimited tax credits to U.S. companies for the removal of CO_2 from the atmosphere at the rate of US$35 per ton of CO_2 removed from the air. From then until the time of writing this book, Global Thermostat has been the only operating U.S. company undertaking DAC. This extraordinary piece of legislation provides unlimited tax credits, for the purpose of funding exactly what the U.N. IPCC says is now needed to avert catastrophic climate change. It is unique and somewhat surprising, as it puts the U.S. at the forefront of global efforts to avert catastrophic climate change.

Since the DAC proposal was written, Rhode Island's Senator Calkin has suggested that Global Thermostat merits a US$100 million Rhode Island state bond to execute what was proposed. Very helpful support and introductions were provided by a Rhode Island Library Manager Wil Gregersen who introduced Global Thermostat and had several discussions with Jeff Tingley, Vice President of the Rhode Island Commerce Corporation, and Bill Ash who oversees the Rhode Island Industrial Facilities Corporation bond issuance (RIIFC) and the Rhode Island Renewable Energy fund — among other Rhode Island state financial services. After a few "nuts and bolts" discussions in March 2018 with Global Thermostat management including financial adviser Neal Cravens and Edgar Bronfman, Chairman of the Global Thermostat board and former CEO of Seagrams, substantial corporate information was provided and Steve Pryor, the Rhode Island Secretary of Commerce, found the Global Thermostat opportunity interesting for the state on account of the possibility to further its Rhode Island Sustainability legislative agenda and also with regards to the region's expansion of jobs and exports.

During this period, Global Thermostat adviser Raymond Chavez, working with Jan Hartke and Ira Magaziner, was introduced in turn to Seth Magaziner, Ira's son and Rhode Island's Treasurer. He expressed interest in the proposal being prepared for a Global Thermostat factory near Providence, Rhode Island. The proposal included, *inter alia*, facilities, perimeters in square feet, numbers of employees to be recruited over time, P&L, how the bond would be utilized, serviced and repaid, and the need for a "draw down" facility that would act as a "line of credit" for Global Thermostat to scale up successfully in the U.S. in the near future. This unexpected and much welcome turn of events demonstrates the real possibilities now opening up to follow the strategies explained in Chapter 9 which can be scaled globally to help reverse climate change. As this book goes to press, a major oil company — one of the largest in the world — expressed a desire to join the carbon negative policies described here and help scale up carbon removals to one gigaton, which is the natural next step. The power is in our hands to make all this a reality and do it. Let's reverse climate change now.

Endnotes

Introduction

1. In a June 1, 2015 open letter to the *Financial Times* Europe's six largest oil and gas groups have united together in seeking help from the United Nations to stop global warming and create a global carbon pricing system. "We owe it to the future generations to seek realistic workable solutions to the challenge of providing more energy while tackling climate change," says the letter. BP, Shell, Eni, Total, Statoil, and BG Group announced in a letter to Christiana Figueres, Chairman of the UNFCCC that they are joining forces for an initiative calling for carbon pricing, which is accomplished through the carbon market emissions-trading systems or taxes.
2. *UNFCCC Newsroom Press Release*, 2015.
3. *IEA*, 2014.
4. Yardley, 2007.
5. McMichael *et al.*, 2003.
6. Ronbine, 2004; *Geophysical Fluid Dynamics Laboratory*, 2015.
7. *UN IPCC*, 2013.
8. Davenport, 2014.
9. *World Resources Institute (WRI)*, 2012.
10. Energy use is expected to increase 5 to 10 times during this century (U.S. Department of Energy). The U.S. coal industry recently presented an energy independence plan to secure subsidies for coal production.
11. *International Energy Agency*, 2008, International Renewable Energy (IRENA).

12. *World Energy Outlook (WEO)*, 2014.
13. *IPCC*, 2013.
14. Downing, 2015.
15. China GDP = $3,865.88 vs. U.S. GDP= $54,360.50 in 2014.
16. *UNEP Synthesis Report*, 2013, Executive Summary p. xi.
17. *WRI*, 2011.
18. *UN IPCC*, 2014.
19. *The World Bank*, 2014.
20. *National Bureau of Statistics of China*, 2015; *United States Census Bureau*, 2015.
21. *Green Peace*, 2010.

Chapter 1

1. Hanson, 2010.
2. The UN FAO's estimate of 18% includes greenhouse gases released in every part of the meat production cycle. Transport, by contrast, accounts for 13% of humankind's greenhouse gas footprint, according to the IPCC. See *UN FAO*, 2006 and *IPCC*.
3. *IPCC*, 2014.
4. *UN IPCC*, 2007.
5. *UNFCCC*, 2000.
6. Hartmann *et al.*, 2013; *World Meteorological Organization (WMO)*, 2014.
7. *National Oceanic and Atmospheric Administration (NOAA)*, 2014.
8. *UN IPCC*, 2007.
9. *Open Geography*, 2015.
10. *NASA*, 2011.
11. Elredge, 2010.
12. *IPCC*, 2014.
13. For full disclosure, one of the authors of this book has co-invented and patented such technology, one that can turn fossil fuel power plants into net sinks of CO_2. This requires carbon capture, but not of the classic variety that scrubs the CO_2 as it is emitted by a power plant. It requires a new form of carbon capture that comes directly from air, Direct Air Capture or DAC. This was until recently thought to be impossible or too expensive to make sense. But the Global

Thermostat technology that is already patented — it has 11 patents, in the U.S. and Japan and by treaty is protected in 147 nations — does exactly that and has sufficiently low costs that it is commercially viable.

14. Poore, 2012.
15. Hanson, 2010.
16. *OECD*, 2007.
17. Pearce, 2005.
18. *Nature Reports*, 2008.
19. *IPCC*, 2013
20. Goldenberg, 2013.
21. *Wildlife Conservation Society*, 2008.
22. Gleik, 2012.
23. Biello, 2014.
24. *IPCC*, 2014.
25. Thompson, 2014.
26. Martin Parry, quoted in Pearce F. 2005.
27. *International Union for Conservation of Nature (IUCN)*, 2014.
28. Kahn, 2015.
29. *NOAA*, 2015.
30. *Science Daily*, 2009.
31. *UNFCCC*, 2014.
32. *UN IPCC*, 2001.
33. Rabatel, 2013.
34. *The World Bank*, 2010.
35. *Ibid.*
36. Fräss-Ehrfeld, 2009; *NRDC*, 2015.
37. Larsen, 2003.
38. *Ibid.*
39. *World Health Organization*, 2003.
40. *UN IPCC*, 2007.
41. Kurz, 2008.
42. Bradshaw and Holzapfel, 2006.
43. *WHO*, 2014.
44. *United Nations Department of Economic and Social Affairs (UNDESA)*, 2014.
45. *Ibid.*
46. *Dhaka Tribune*, 2013.
47. *UN IPCC*, 2007.

48. Death tolls are extraordinarily high as a result, natural disasters responsible for 9,700 deaths each year on average over the last 10 years according to Swiss Re. In 2007, natural catastrophes claimed 15,000 lives worldwide, 11,000 in Asia alone. Cyclone Sidr caused more than 3,300 deaths in Bangladesh as reported by UNICEF. As shown by Majra J. and A. Gur in "Climate change and health: Why should India be concerned?" weather-related natural disasters killed 600 thousand people worldwide in the 1990s. Of those deaths, 95% were in poor countries.
49. *UN IPCC*, 2007.
50. *EIA*, 2012.
51. *Ibid.*; OECD, 2015.
52. Pope Francis, 2015.
53. Martin Parry, quoted in Pearce, 2005.
54. Stern, 2006.
55. Hanson, 2010.
56. Harvey, 2013.
57. Pearce, 2014. Referencing *IPCC*, 2013.
58. For full disclosure, one of the authors of this book has co-invented and patented such technology, one that can turn fossil fuel power plants into net sinks of CO_2. This requires carbon capture but not of the classic variety that scrubs the CO_2 as it is emitted by a power plant. It requires a new form of carbon capture that comes directly from air, Direct Air Capture or DAC. This was thought until recently to be impossible or too expensive to make sense. But the Global Thermostat technology that is already patented — it has 61 patents, in the U.S. and Japan and by treaty is protected in 147 nations — does exactly that and has sufficiently low costs that it is commercially viable.

Chapter 2

1. *UN*, 1999.
2. *Ibid.*
3. *The Economist*, 2013.
4. *EIA*, 2014.
5. Perry, 2013.
6. *UNEP*, 2015.

7. https://www.blackrock.com/corporate/investor-relations/larry-fink-ceo-letter.
8. https://www.cnbc.com/2019/12/19/goldman-pledges-750-billion-for-opportunities-in-sustainable-finance.html.
9. https://www.cnbc.com/2020/01/24/climate-crisis-investors-putting-money-towards-sustainable-future.html.
10. https://www.cnbc.com/2020/01/24/climate-crisis-investors-putting-money-towards-sustainable-future.html.
11. https://www.cnbc.com/2020/01/14/esg-funds-see-record-inflows-in-2019.html.
12. https://www.cnbc.com/2020/01/14/esg-funds-see-record-inflows-in-2019.html.
13. https://www.oilandgas360.com/eu-wants-to-spend-1-trillion-to-help-make-it-climate-neutral-by-2050/.
14. Carbon concentration in the atmosphere is a "public good" as defined by economists because it is the same all over the world. The first time climate change was identified with a public good was in Chichilnisky and Heal, 1994.
15. *World Resource Institute (WRI)*, 2014.
16. *CNN*, 2015.
17. DeCanio, 2009.
18. *Global Greenhouse Warming*, 2015.
19. Ackerman *et al.*, 2009; Ackerman and Heinzerling, 2004.
20. Ramsey, 1928. See also Ackerman *et al.*, 2009; Heal, 2000.
21. Ackerman and Heinzerling, 2004.
22. Chichilnisky, 1993, 1996, 2000, 2009 and 2015.
23. Bromley, 1992.
24. Chichilnisky and Heal, 1993.
25. Risks that are created by our own actions were introduced to the economic literature by Chichilnisky, 1996(a), an article that received the Leif Johansen's award at the University of Oslo.
26. Ackerman *et al.*, 2009.
27. *UN IPCC*, 2014.
28. Stern, 2006.
29. *The World Bank*, 2015.
30. *U.S. Department of the Treasury*, 2008.
31. Trotta, 2015.
32. Stern, 2006.

33. Ackerman and Stanton, 2008.
34. *OECD*, 2007.
35. *DARA*, 2012.
36. Iverson, 2013.
37. *Hurricanes: Science and Society*, 2015.
38. Swiss Re Economic Research and Consulting, Sigma No. 1/2014.
39. Swiss Re, Sigma No. 3/2008.
40. *Sigma Explorer*, 2015.
41. *Swiss Re Economic Research and Consulting*, 2014.
42. Swiss Re Economic Research and Consulting, Sigma No. 1/2008.
43. Stern, 2006; *IPCC*, 2007.
44. Chichilnisky, 2009(a); World Bank, 2006, 2007.
45. Chestney, 2015; *IMF World Economic Outlook (WEO)*, 2015; *EIA*, 2015.
46. Chichilnisky and Eisenberger, 2009; Eisenberger and Chichilnisky, 2007.
47. Hassol, 2011.
48. *EIA*, 2014.
49. Wind, biomass, hydroelectric, solar, geothermal, nuclear, and possibly fusion.
50. By the end of this century, it is expected to be 5 to 10 times today's energy use. US DOE.
51. *Environment America Research and Policy Center*, 2014.
52. 89% of the energy used today comes from fossil fuels; less than 1% from renewable sources; 0.01% is solar energy.
53. Cohen *et al.*, 2009 and Eisenberger and Chichilnisky, 2007; Chichilnisky and Eisenberger, 2009.
54. Scientists consider the possibility of a "tipping point," a level of heating that triggers catastrophic climate change, which is typical of physical systems with complex feedback effects. The earth's climate is generally believed to be one of them. In general, one views the risks as having "heavy tails" so rare events turn out to be more frequent than usually expected.
55. Transitioning away from fossil fuels in a short period of time could lead to social disruption, since most human life is dependent on energy.
56. Chichilnisky and Eisenberger, 2007 and 2009, US DOE.

57. Jones, 2008, 2009; Eisenberger and Chichilnisky, 2007; Chichilnisky and Eisenberger, 2009. Simultaneous production of electricity and air capture is called "cogeneration."
58. Cohen *et al.*, 2009; Chichilnisky, 2008(a).
59. Eisenberger and Chichilnisky, 2007; Chichilnisky and Eisenberger, 2009.
60. Jiang, 2013.

Chapter 3

1. GDP growth (annual %). *The World Bank*, 2014.
2. *UNEP*, 2013.
3. Tyndall, 1861.
4. *National Drought Policy Commission Report*, 2000.
5. Hansen Senate Testimony, 1988.
6. Chichilnisky, 1977(a) and 1977(b).
7. Article 2, p. 9, UNFCCC, 1992.
8. *UNFCCC*, 1992.
9. *EPA*, 2015.
10. Chichilnisky, 1977, 1994(a), 1994(b), 1995, 1996 and 1997; Chichilnisky and Heal, 1993, 1994, 1995 and 2000.
11. Chichilnisky and Heal, 2000.
12. Chichilnisky and Heal, 1995; Chichilnisky, 1994(b).
13. Chichilnisky, 1997.
14. See Chichilnisky Pegram Lectures on www.chichilnisky.com "Books and writings."
15. Cap-and-trade systems cap global emissions and bind the emissions rights of polluters. Awarding emissions rights on the basis of past emissions levels is often called "grandfathering."
16. Hourcade and Ghersi, 2002. The CDM allows credits for industrial nations' projects that are carried out on developing nations' soil and are proven to reduce emissions. These can be traded in the carbon market. This was my preferred approach to how to integrate developing nations into the Kyoto Protocol carbon market while still abiding by the Article 4 of the UNFCCC, which does not allow limits on developing nations' emissions without compensation. I ran a conference on this topic at Columbia University with the negotiators of the Kyoto Protocol

in 1997–1998: "From Kyoto to Buenos Aires: Technology Transfer and Carbon Trading." Ambassadors Kilaparti Ramakrishna and Raul Estrada Oyuela participated as main speakers, among others.

17. *National Snow & Ice Data Center*, 2009.
18. Geman, 2015.

Chapter 4

1. Bart Chilton, quoted by *Reuters News*, June 25, 2008.
2. Chichilnisky, 2009(b); Sheeran 2006(a); Chichilnisky and Heal, 1994, 2000.
3. Adam Smith described how market forces — the invisible hand — can harness individual self-interest for the greater social good. See Smith, 1776.
4. *UNFCCC*, 2015.
5. Cohen T. *et al.*, 2014.
6. Hulac, 2015; Lund, H. *et al.*, 2015.
7. Pope Francis, 2015.
8. *Silicon Valley Index*, 2014.
9. *UNEP*, 2015, p. 2; *Bloomberg*, 2013.
10. Gupta, 2012.
11. Hone, 2015.
12. *IEA*, 2008.
13. *World Bank Group*, 2015.
14. Estrada Oyuela, 2000. The climatic last minute addition was the carbon market and its attendant CDM.
15. Article 2, UNFCCC, 1992.
16. Countries were to reduce their emissions to 1990 levels by 2000. Since 1990, emissions in the U.S. have actually increased by more than 15%.
17. Estrada Oyuela, 2000.
18. Article 3, The Kyoto Protocol to the UNFCCC, 1997.
19. The Kyoto Protocol CDM provides incentives to curb deforestation in developing countries under certain conditions.
20. Article 2.3. See Estrada Oyuela, 2000.
21. Chichilnisky and Heal, 1994, 2000; Sheeran, 2006. This is not true for sulfur dioxide.

22. This was the rationale proposed by Chichilnisky in her introduction of a carbon market with a preferential treatment for developing nations. See Chichilnisky, 1997; Chichilnisky and Heal, 1994 and 1995.
23. Chichilnisky and Heal, 1995.
24. *Ibid.*
25. *OECD*, 2015.
26. Table 1 "Annex I Emissions Target" provides a breakdown for industrial and developing nations.
27. *OECD*, 2015.
28. Gerber *et al.*, 2013.
29. Chichilnisky and Heal, 1994, 1995 and 2000; Chichilnisky, 2009(b) and 1994(a).
30. GDP per capita (current US$). *The World Bank*, 2014.
31. Chichilnisky and Heal, 1994, 1995 and 2000.
32. *UN Climate Change Newsroom*, 2015.
33. *World Bank Group*, 2012.
34. *World Bank Group*, 2012, p. 44, 45.
35. Chichilnisky, 2009(c).
36. Estrada Oyuela, 2000.
37. *UNFCCC*, 2015.
38. Estrada Oyuela, 2000. The Kyoto Protocol does not limit the amount of purchased carbon reduction credits a country can use to meet its emissions cap. It does, however, state that: "trading shall be supplemental to domestic actions for the purpose of meeting quantified emission limitation and reduction commitments." There are further restrictions on the use of specific types of CDM project as well.
39. Chichilnisky and Heal, 1994, 1995 and 2000.
40. *World Bank Group*, 2012.
41. Annex I countries are essentially the rich nations of the Organization for Economic Cooperation & Development Paris, France, or OECD.
42. *UNFCCC*, 2015.
43. *World Bank Group*, 2012.
44. *World Bank Group*, 2015.
45. *World Bank Group*, 2012.
46. *Ibid.*, p. 44, 45.
47. *Ibid.*

48. Hone, 2015.
49. *World Bank Group*, p. 1.
50. *UNFCCC*, 2012.

Chapter 5

1. *UNFCCC*, 2014.
2. *The United Nations Regional Information Centre for Western Europe Magazine*, Issue No. 16, December 2007.
3. Chichilnisky presented this proposal at the following: the Bali COP on December 13, 2007; the International Monetary Fund in July and October 2007; a Bipartisan Briefing in Capitol Hill organized at U.S. Congress by Rep. Michael Honda in May 2008, with the participation and support of seven Members of the House; at the Parliament in Victoria, Australia, in November 2008; at a Caucus on Sustainable Energy and the Environment in Capitol Hill, with the participation of 42 Members of the U.S. House of Representatives, on March 31, 2009; and at the UNCTAD Expert Meeting on Trade and Investment Opportunities and Challenges under the Clean Development Mechanism in Geneva, April 27–29, 2009.
4. Campbell, 2008.
5. Related to "clean coal" proposals to produce electricity from coal in a carbon neutral way through CCS technologies. Available at: www.nma.org/ccs/aboutccs.asp. See U.S. DOE information on ongoing projects.
6. Chichilnisky and Eisenberger, 2009; Eisenberger and Chichilnisky, 2007; Jones, 2008, 2009; Chichilnisky, 2008(a), and Cohen *et al.*, 2008.
7. *UNFCCC*, 2014.
8. *World Bank Group*, p. 10.
9. *UNFCCC*, 2012.
10. Tollefson, 2008.
11. *World Bank*, 2006, 2007.
12. Tollefson, 2008.
13. *BBC News*, 2008.
14. Quote from Arthur Runge-Metzger who oversees climate issues at the EC: "The EC needs to spur new technologies now, because paying for offsets elsewhere won't solve the

problem," quoted in Tollefson, 2008. Paying for offsets elsewhere (in developing nations) is the nature of the CDM.

15. Miles Austin, head of the European regulatory affairs for carbon offset dealer Ecosecurities, Dublin.
16. Zhang, 2010.
17. *OECD*, 2007.
18. *UN IPCC*, 2014.
19. *The Economist*, November 15, 2007.
20. Buckley, 2008.
21. On November 12, 2008, Chichilnisky addressed Members of Parliament in Melbourne, Australia on the commercial benefits of the carbon market.
22. *The World Bank, 2015.*
23. *IPCC*, 2007.
24. China's GDP is approximately $4000.
25. Chichilnisky, 2009(a).
26. *Energy Information Agency (EIA)*, 2014.
27. *World Resource Institute (WRI)*, 2012.
28. Milanovic, 2006.
29. GDP per capita (current US$). *The World Bank*, 2015.
30. *WRI*, 2012.; GDP (current US$). *The World Bank*, 2015.
31. Zhang, 1999.
32. Biagini, 2000.
33. Buckley, 2008.
34. Traditional Chinese medicine would diagnose our planet's health as suffering of "Too much Fire".
35. Chichilnisky, 1994(a).
36. *Ibid.*
37. Chichilnisky, 1981.
38. Chichilnisky, 1981; Rodrick, 2006.
39. Chichilnisky, 1995.
40. Chichilnisky, 1994(a).
41. *U.S. Census Bureau*, 2015.

Chapter 6

1. Ackerman and Stanton, 2006; Stern, 2006.
2. *World Bank*, 2006, 2007.
3. Chestney, 2015.

4. For the critical timing issue see the 2006 Stern Review.
5. *WRI*, 2012.
6. GDP (current US$). *The World Bank*, 2014.
7. Dargin, 2013.
8. Boyce and Riddle, 2007, as well as Office of Congressional Research and other scholarly works.
9. *UNFCCC*, 2012.
10. *UNFCCC*, 2015(a).
11. *Ibid.*
12. *Ibid.*
13. Project descriptions at *UNFCCC*, 2015(b).
14. Chichilnisky and Heal, 1994, 1995 and 2000; Sheeran, 2006(a) and 2006(b).

Chapter 7

1. Chichilnisky and Wu, 2006 predicted exactly this phenomenon a year before it started. See also Chichilnisky, 2008a and 2008b.
2. Chichilnisky and Wu, 2006 provides a rigorous demonstration of this proposition.
3. Oak, 2008.
4. *International Energy Agency*, 2008.
5. Chichilnisky is working in a commercial demonstration plant for one of the negative carbon technologies available today, with physicists and business people in the U.S.
6. *UN Millennium Report*, 2000.
7. Chichilnisky, 1998.
8. *Ibid.*
9. *Ibid.*
10. *Ibid.*
11. *Ibid.*
12. *Food and Agriculture Organization of the United Nations (FAO)*, 2010.
13. *Trucost*, 2013.
14. Gewirtz, 2009.
15. Chichilnisky, 2009(a).
16. Increased drilling in Alaska will not have a perceptible impact on global oil prices since the resulting increase in supply from drilling will be small relative to current world supplies.

17. Ackerman *et al.*, 2009.
18. *Ibid.*
19. Boyce and Riddle, 2007.
20. Ackerman *et al.*, 2009.

Chapter 8

1. *Cf.* Hyun Youk and Alexander van Oudenaarden, *Nature*, 2010.
2. Chichilnisky "The Gender Gap" in 2014.
3. Research shows that men are admired for traits that prevail in negotiating salaries, while the same traits are considered too aggressive for females.
4. *The Times*, 2011.
5. Chichilnisky, 1995 and 1998.
6. Chichilnisky, 1994.
7. Chichilnisky, 1994.
8. Chichilnisky and Heal, 1994
9. Chichilnisky, *Time Magazine*, 2009.
10. *Cf.* Chichilnisky and Sheeran, 2009.
11. This is according to Nick Stern (2006).
12. The World Bank, "Status and Trends of the Carbon Market", 2005–2010.
13. See Chichilnisky, 2009.
14. Chichilnisky, 1977a, and 1977b; and Herrera *et al.*, 1976.
15. Grundtland Report, 1992.
16. Chichilnisky, 1993.
17. The basis for Green capitalism was explained in *Time Magazine* (Chichilnisky, 2009).
18. See *The Economist* issue on "The Prosperity Puzzle", April 30, 2016, p. 10, and "The Modern Economy", p. 7
19. *IPCC*, Fifth Assessment Report, p. 191.
20. Chichilnisky and Eisenberger, 2011.
21. e. g. Chichilnisky and Eisenberger, 2011.
22. Including "World's Top Ten Most Innovative Company" in Energy from Fast Company Magazine, and "World's Top 50 Innovator in Renewable Energy."
23. For example, Global Thermostat (www.globalthermostat.com).

24. Includes General Electric, SSE, Siemens, Linde, as well as new and smaller firms.
25. See The World Bank's "Status and Trends of the Carbon Market", 2010 and 2011.
26. *The Economist*, "The Trouble with GDP", p. 21, April 30, 2016.
27. Ibid, p. 22.
28. Chichilnisky and Eisenberger, 2012.
29. Chapter 6 of this book.
30. For example, a recent BMW automobile model uses only carbon fibers rather than metals.

Chapter 9

1. European Commission's webpage on the event. Available at: http://ec.europa.eu/clima/policies/international/negotiations/paris/index_en.htm.
2. Ibid.
3. Ibid.
4. The UNFCCC Kyoto Protocol history is in Chichilnisky and Sheeran (2010).
5. http://www.greenbuildermedia.com/blog/china-and-india-doing-the-unimaginable-to-address-climate-change
6. Experts such as NASA climate scientist Professor James Hansen said that the agreement is fraud and a fake, "just worthless words. There is no action, just promises", *The Guardian*. Available at: http://www.theguardian.com/environment/2015/dec/12/james-hansen-climate-change-paris-talks-fraud.
   George Monbiot writes of the Paris Agreement "By comparison to what it could have been, it is a miracle. By comparison to what it should have been, it is a disaster", *The Guardian*. Available at: http://www.theguardian.com/environment/georgemonbiot/2015/dec/12/paris-climate-deal-government-fossil-fuels.
7. Chichilnisky, G. and C. J. Polychroniou, Paris COP21 climate agreement is bound to nothing: What is the solution? *E-International Relations*, 2016. Available at: http://www.e-ir.info/2015/12/29/paris-cop21-climate-agreement-is-bound-to-nothing-what-is-the-solution/.
8. Chichilnisky proposed and trademarked this term in 2009.
9. See e.g. www.globalthermostat.com.

10. SRI is located at 333 Ravenswood Avenue, Menlo Park, California and accepts visitors.

Chapter 10

1. (i) Global Warming of 1.5C. IPCC October 2018. https://report.ipcc.ch/sr15/pdf/sr15_spm_final_pdf.
   (ii) Negative Emissions Technologies and Reliable Sequestration: A Research Agenda, US National Academy of Sciences — this title will be released on July 24th 2019. Highlights available at https://www.nap.edu/resource/25259/Negative%20Emissions%20Technologies.pdf.
   (iii) Fourth National Climate Assessment Volume II Impacts, Risks, and Adaptation in the United States, Report-in-brief US Global Change Research Program October 2018, https://nca2018.globalchange.gov/downloads/NCA4_2018_FullReport.pdf
   (iv) Scientists Push for a Crash Program to Scrub Carbon From the Air, New York Times Brad Plumer October 24th 2018, https://www.nytimes.com/2018/10/24/climate/global-warming-carbon-removal.html.
2. Negative Emissions Technologies and Reliable Sequestration. Consensus Study Report, National Academies of Sciences USA, October 2018.
3. Fast Company Magazine selected Global Thermostat "World's Top 10 Most Innovative Companies" in energy in April 2015. Chichilnisky was selected as the CEO of the Year in 2015, and more recently in the February 2019 MIT Technology Review — curated by Bill Gates — designated Global Thermostat technology a "Top Ten Breakthrough Technologies 2019."
4. GT's technology has been designed to be resistant to NOx & SOx levels found in most flues. Where near-zero NOx/SOx levels are required in GT's output CO^2 (i.e. for food-grade uses), we can either upgrade our output, or capture directly from air, avoiding the presence of any meaningful levels of contaminants from the start.
5. Other cost drivers include: energy to move the process stream, electricity to power blower fans to move air through the monoliths; and capital cost. The upfront and replacement cost of the sorbent also contributes to the overall cost performance.

6. Detailed third-party reports are available, and have been produced by:

    - **2015 Sargent & Lundy FEED:** Study conducted on behalf of GT & NRG Energy (see APPENDIX I)
    - **Det Norsk Veritas (DNV):** premier global risk & technology assessment firm. Two technology reports: 2011 & 2013. www.dnvusa.com
    - **SRI International:** leading US technology development institute, specializing in carbon capture. (www.sri.com)
    - **Georgia Institute of Technology:** a leading US carbon capture academic research center. www.gatech.edu/research/institute
    - **U.S. National Academy of Sciences (NAS):** A United States nongovernmental organization and part of the National Academies of Sciences, Engineering, and medicine. http://www.nasonline.org/about-nas/mission/

7. The article "Carbon Negative Power Plants" was authored by G. Chichilnisky and P. Eisenberger and it appeared in *CryoGas International* in 2013: https://www.gasworld.com/carbon-negative-power-plants/2005090.article

Bibliography

53,000 die from malnutrition every year in Bangladesh. *Dhaka Tribune*, December 19, 2013. Available at: http://www.dhakatribune.com/health/2013/dec/19/53000-die-malnutrition-every-year-bangladesh.

Ackerman, F., S.J. DeCanio, J. Stephen, R. Howarth and K.A. Sheeran. The limitations of integrated assessment models of climate change. *Climatic Change*, 95(3), 2009, p. 297–315.

Ackerman, F. and E. Stanton. *The Costs of Climate Change: What We'll Pay if Climate Change Continues Unchecked*. NRDC, 2008. Available at: http://www.nrdc.org/globalwarming /cost/fcost.pdf.

Ackerman, F. and E. Stanton. Climate change — the costs of inaction. Of Friends of the Earth, 2006. Available at: http://www.foe.co.uk/resource/reports/econ_costs_cc.pdf.

Ackerman, F. and L. Heinzerling. *Priceless*. The New Press, 2004.

Agricultural Output. *OECD*, 2015. Available at: https://data.oecd.org/agroutput/meat-consumption.htm.

Baumert, K.A. and N. Kete. *Will Developing Countries Carbon Emissions Swamp Global Emissions Reduction Efforts?* World Resources Institute, 2002.

Benefits of the Clean Development Mechanism 2012. *UNFCCC*, 2012, p. 8.

Biagini, B. (ed.). *Confronting Climate Change: Economic Priorities and Climate Protection in Developing Nations*. National Environmental Trust, Washington DC, 2000.

Biello, D. Biodiversity survives extinction for now. *Scientific American*, April 20, 2014. Available at: http://www.scientificamerican.com/podcast/episode/biodiversity-survives-extinctions-for-now1/.

Black, R. Nature loss "dwarfs bank crisis." *BBC News*, October 10, 2008.

Boyce, J. and M. Riddle. *Cap and Rebate: How to Curb Global Warming while Protecting the Incomes of American Families*, Political Economy Research Institute Working Paper No. 150, 2007.

Boyds Forest Dragon. *Global Greenhouse Warming*, 2015. Available at: http://www.global-greenhouse-warming.com/Boyds-Forest-Dragon.html.

Bradshaw, W.E. and C. Holzapfel. Perspectives Section, Science, 312, June 2006, p. 1477, 1478.

Bromley, D.W. *Making the Commons Work*. ICS Press, San Francisco, 1992.

Buckley, C. China report warns of greenhouse gas leap. *Reuters News Service*, October 22, 2008.

CAIT Climate Data Explorer. *WRI*, 2011.

CAIT — Country Greenhouse Gas Emissions Data. Climate Data Explorer. *WRI*, 2012.

Carbon Emissions (kT). *The World Bank*, 2014.

Campbell, W. *Reducing Carbon Capture and Storage: Assessing the Economics*. McKinsey & Company, 2008. Available at: http://www.mckinsey.com/clientservice/ccsi/pdf/CCS_Assessing_the _Economics.pdf.

Carbon is forever. *Nature Reports*, November 20, 2008. Available at: http://www.nature.com/climate/2008/0812/full/climate.2008.122.html.

Carbon Pricing Watch 2015: An Advance Brief from the State and Trends of Carbon Pricing 2015 Report, to be Released Late 2015. *World Bank Group*, 2015, p. 1, 10.

CDM Insights. *UNFCCC*, 2015(a).

Chestney, N. EU carbon market expects price rise for first time in four years. *Reuters*, May 26, 2015.

Chichilnisky, G. Biodiversity as knowledge. *Proceedings of the National Academy of Sciences*, 1995.

Chichilnisky, G. (a). *Beyond the Global Divide: From Basic Needs to the Knowledge Revolution*. 2009.

Chichilnisky, G. (ed.) (b). *The Economics of Climate Change*. Edward Elgar, Library of Critical Writings in Economics, 2009.

Chichilnisky, G. (c). Le paradoxe des Marches verts. *Les Echos*, January 21, 2009. Available at: http://www.lesechos.fr/info/analyses/4822817 -le-paradoxe-des-marches-verts.htm.

Chichilnisky, G. (a). Energy security, economic development and climate change: Short and Long Term Challenges. El Boletin Informativo Techint No. 325, April 2008, pp. 53–76.

Chichilnisky, G. (b). How to restore the stability and health of the economy. *Huffington Post*, 2008. Available at: http://www.huffington post.com/graciela- chichilnisky/its-the-ortgageshow-to_b_144376.html.

Chichilnisky, G. Economics returns from the biosphere. *Nature*, 391, 1998, p. 629–630.

Chichilnisky, G. *Development and Global Finance: The Case for an International Bank for Environmental Settlements, Report No. 10.* United Nations Development Program and the United Nations Educational Scientific and Cultural Organization, May 1997.

Chichilnisky, G. Forward trading: A proposal to End the Stalemate between the US and China on Climate Change. *Time Magazine*, Special Issue of Heroes of the Environment, last page, December 2009.

Chichilnisky, G. (a). Markets with endogenous uncertainty: Theory and policy. *Theory and Decision*, 41, 1996, p. 91–131.

Chichilnisky, G. (b). The greening of Bretton Woods. *Financial Times*, January 10, 1996.

Chichilnisky, G. The economic value of the earth's resources. Invited Perspectives Article, *Trends in Ecology and Evolution (TREE)*, 1996, p. 135–140.

Chichilnisky, G. (a). North–South trade and the global environment. *American Economic Review*, 84(4), 1994, p. 851–874.

Chichilnisky, G. (b). The trading of carbon emissions in industrial and developing nations. In: Jones (ed.) *The Economics of Climate Change*, OECD, Paris, 1994.

Chichilnisky, G. *What is Sustainable Development?* Paper presented at the 1993 workshop of the Stanford Institute for Theoretical Economics, 1993.

Chichilnisky, G. Terms of trade and domestic distribution: Export led growth with abundant labor. *Journal of Development Economics*, 8, 1981, p. 163–192.

Chichilnisky, G. (a). Economic development and efficiency criteria in the satisfaction of basic needs. *Applied Mathematical Modeling*, 1(6), 1977, p. 290–297.

Chichilnisky, G. (b). Development patterns and the international order. *Journal of International Affairs*, 1(2), 1977, p. 274–304.

Chichilnisky, G. and P. Eisenberger. Energy security, economic development and global warming: addressing short and long term challenges. In: G. Chichilnisky (ed.) *The Economics of Climate Change*, Edward Elgar, Cheltenham, 2009.

G. Chichilnisky and P. Eisenberger. Carbon negative power plants. *Cryogas International,* 2011

Chichilnisky, G. and G. Heal. *Environmental Markets: Equity and Efficiency*. Columbia University Press, 2000.

Chichilnisky, G. and G. Heal. *Markets for Tradeable CO2 Emissions Quotas: Principles and Practice, OECD Report No. 153*. OECD, Paris, 1995.

Chichilnisky, G. and G. Heal. Who should abate carbon emissions: An international perspective. *Economic Letters*, Spring, 1994, p. 443–449.

Chichilnisky, G. and G. Heal. Global environmental risks. *Journal of Economic Perspectives*, 7(4), 1993, p. 65– 86.

Chichilnisky, G. and K. Sheeran. *Saving Tokyo*. New Holland, 2010.

Chichilnisky, G. and H.M. Wu. General equilibrium with endogenous uncertainty and default. *Journal of Mathematical Economics*, 42, 2006, p. 499–524.

Chichilnisky, G. The economic value of the earth resources. In: E. Futter (ed.) *Scientists on Biodiversity*. American Museum of Natural History, New York, 1998.

Climate Change and Health. *WHO*. August 2014. Available at: http://www.who.int/mediacentre/factsheets/fs266/en/.

Climate Change and Human Health — Risks and Responses. Summary. *World Health Organization (WHO)*, 2003. Available at: http://www.who.int/globalchange/summary/en/index10.html.

Climate Change Indicators in the United States. *EPA*, 2015. Available at: http://www.epa.gov/climatechange/science/indicators/weather-climate/temperature.html.

Climate Change Information Sheet 11. *UNFCCC*, 2014. Available at: http://unfccc.int/essential_background/background_publications_htmlpdf/climate_change_information_kit/items/290.php.

Climate Vulnerability Monitor. *DARA*, 2012. Available at: http://daraint.org/climate-vulnerability-monitor/climate-vulnerability-monitor-2012/.

Cohen, R., G. Chance, G. Chichilnisky, P. Eisenberger and N. Eisenberger. Global Warming and Carbon-Negative Technology: Prospects for a Lower-Cost Route to a Lower-Risk Atmosphere *Energy & Environment*, 2009. Available at: http://dx.doi.org/10.2139/ssrn.1522281.

Cohen T. *et al.* Obama's big environmental move: Power plants to cut carbon pollution. *CNN*, September 8, 2014. Available at: http://www.cnn.com/2014/06/02/politics/epa-carbon-emissions/.

Dargin, J. Why a global carbon market is coming sooner than you think. *Fair Observer*, January 14, 2013.

Davenport, C. Pentagon signals security risks of climate change. *New York Times*, October 13, 2014. Available at: http://www.nytimes.com/2014/10/14/us/pentagon-says-global-warming-presents-immediate-security-threat.html.

DeCanio, S.J. The political economy of global carbon emissions reduction. *Ecological Economics*, 68, 2009, p. 915–924.

Development and Climate Change. Main Messages of the World Development Report 2010. *The World Bank*, 2010, p. xx.

Distribution of Registered Projects by Host Party. *UNFCCC*, June 30, 2015. Available at: https://cdm.unfccc.int/Statistics/Public/files/201506/proj_reg_byHost.pdf.

Downing, L. Clean energy investment jumps 16%, shaking off oil's drop. *Bloomberg*, January 9, 2015.

Do We Have Enough Oil Worldwide to Meet Our Future Needs? *EIA*, 2014. Available at: http://www.eia.gov/tools/faqs/faq.cfm?id=38&t=6.

Eisenberger, P. and G. Chichilnisky. *Reducing the Risk of Climate Change While Producing Renewable Energy*. Columbia University, 2007.

Elredge, S. and B. Beik. Ice Ages — What are they and what Causes Them? *Utah Geological Survey*, September 3, 2010. Available at: http://geology.utah.gov/map-pub/survey-notes/

glad-you-asked/ice-ages-what-are-they-and-what-causes-them/.

Environment. *Energy Information Agency (EIA)*, 2014. Available at: http://www.eia.doe.gov/environment.html.

Estrada Oyuela, R. A commentary on the Kyoto Protocol. In: G. Chichilnisky and G. Heal (eds.) *Environmental Markets: Equity and Efficiency*, Columbia University Press, New York, 2000.

Fräss-Ehrfeld, C. *Renewable Energy Sources: A Chance to Combat Climate Change.* Kluwer Law International, 2009, p. 13.

From Despair to Repair: Dramatic Decline of Caribbean Corals can be Reversed. *International Union for Conservation of Nature (IUCN)*, 2014. Available at: http://www.iucn.org/?16050/1/From-despair-to-repair-Dramatic-decline-of-Caribbean-corals-can-be-reversed.

G-20 Leaders' Statement from Meeting in Pittsburgh, USA, September 24–25, 2009. Available at: http://www.pittsburghsummit.gov/mediacenter/129639.htm.

Geman, B. Big oil companies want a price on carbon. Here's why. *National Journal*, 2015. Available at: http://www.nationaljournal.com/energy/climate-change-fracking-paris-bp-shell-20150601.

Gerber, P.J., H. Steinfeld, B. Henderson, A. Mottet, C. Opio, J. Dijkman, A. Falcucci and G. Tempio. *Tackling Climate Change Through Livestock — A Global Assessment of Emissions and Mitigation Opportunities.* Food and Agriculture Organization of the United Nations (FAO), Rome, 2013.

Gewirtz, D. The dot-com bubble: How to lose $5 trillion. *CNN*, November 24, 2009.

Gupta, R. The Global Energy Observatory: A One Stop Site for Information on Global Energy Systems. *Global Energy Observatory*, 2012. Available at: http://globalenergyobservatory.org/docs/analysis_papers/GEO_laThuile.pdf.

Gleik, P. The world's water. *The Biennial Report on Freshwater Resources*, 7, 2012, p. 105.

Global Coal Risk Assessment: Data Analysis and Market Research. World Resources Institute (WRI), November 2012.

Global Trends in Renewable Energy Investment. *UNEP*, 2015, p. 15. Available at: http://fs-unep-centre.org/sites/default/files/attachments/key_findings.pdf.

Global Warming and Hurricanes. *Geophysical Fluid Dynamics Laboratory*, June 15, 2015.

Goldenberg, S. Relocation of Alaska's sinking Newtok village halted. *The Guardian*, August 5, 2013. Available at: http://www.theguardian.com/environment/2013/aug/05/alaska-newtok-climate-change.

Greenland ice sheet melting faster than expected; larger contributor to sea-level rise than thought. *Science Daily*, June 13, 2009. Available at: www.sciencedaily.com/releases/2009/06/090612092741.htm.

Hansen Senate Testimony, 1988. Available at: http://climatechange.procon.org/sourcefiles/1988_Hansen_Senate_Testimony.pdf.

Hanson, S. *et al.* A global ranking of port cities with high exposure to climate extremes. *Climatic Change: Cities and Climate Change Special Issue*, Springer, Netherlands, 2010.

Hartmann, D.L., A.M.G. Klein Tank, M. Rusticucci, L.V. Alexander, S. Brönnimann, Y. Charabi, F.J. Dentener, E.J. Dlugokencky, D.R. Easterling, A. Kaplan, B.J. Soden, P.W. Thorne, M. Wild and P.M. Zhai. Observations: Atmosphere and Surface. In: T.F. Stocker, D. Qin, G.-K. Plattner, M. Tignor, S.K. Allen, J. Boschung, A. Nauels, Y. Xia, V. Bex and P.M. Midgley (eds.) *Climate Change 2013: The Physical Science Basis. Contribution of Working Group I to the Fifth Assessment Report of the Intergovernmental Panel on Climate Change.* Cambridge University Press, Cambridge, United Kingdom and New York, NY, USA. 2013, p. 161.

Harvey, F. IPCC: 30 years to climate calamity if we carry on blowing the carbon budget. *The Guardian*, September 27, 2013. Available at: http://www.theguardian.com/environment/2013/sep/27/ipcc-world-dangerous-climate-change.

Hassol, S.J. Emissions Reductions Needed to Stabilize Climate. *Presidential Climate Action Project*, 2011. Available at: https://www.climatecommunication.org/wp-content/uploads/2011/08/presidentialaction.pdf.

Heal, G. *Valuing the Future*. Columbia University Press, 2000.

Herrera, A., H. D. Scolnik, G. Chichilnisky, G. C. Gallopin, J. E. Hardoy, D. Mosovich, E. Oteiza, G. L. de Romero Brest, C. E. Suirez and L. Talavera. *Catastrophe or New Society: A Latin*

American World Model. International Development Research Centre, Ottawa Canada, 1976.

Hone, D. Killing carbon: Ten years of the EU ETS. *Business Spectator*, January 20, 2015.

Hourcade, J.C. and F. Ghersi. The economics of a lost deal: Kyoto — The Hague — Marrakesh. *The Energy Journal*, 23(3), 2002, p. 1–26.

How Much will Earth Warm if Carbon Dioxide Doubles Pre-Industrial Levels. *National Oceanic and Atmospheric Administration (NOAA)*, January 24, 2014. Available at: https://www.climate.gov/news-features/climate-qa/how-much-will-earth-warm-if-carbon-dioxide-doubles-pre-industrial-levels.

Hulac, B. *Oil and Gas CEOs Call for Carbon Price as Exxon, Chevron Outline Climate Strategy*. E&E Publishing, 2015. Available at: http://www.eenews.net/stories/1060019472.

International Decade for Action 'Water for Life' 2005–2015. *United Nations Department of Economic and Social Affairs (UNDESA)*, 2014. Available at: http://www.un.org/waterforlifedecade/scarcity.shtml.

International Energy Agency (IEA). *Carbon Dioxide Capture and Storage: A Key Carbon Abatement Option*. IEA, 2008.

International Report Confirms: 2014 was Earth's Warmest Year on Record. *NOAA*, July 16, 2015. Available at: http://www.noaanews.noaa.gov/stories2015/071615-international-report-confirms-2014-was-earths-warmest-year-on-record.html.

IPCC. Summary for policymakers. In: T.F. Stocker, D. Qin, G.-K. Plattner, M. Tignor, S.K. Allen, J. Boschung, A. Nauels, Y. Xia, V. Bex and P.M. Midgley (eds.) *Climate Change 2013: The Physical Science Basis*. Contribution of Working Group I to the Fifth Assessment Report of the Intergovernmental Panel on Climate Change, 2013.

IPCC. Climate change 2014: Impacts, adaptation, and vulnerability. In: C.B. Field, C.B., V.R. Barros, D.J. Dokken, K.J. Mach, M.D. Mastrandrea, T.E. Bilir, M. Chatterjee, K.L. Ebi, Y.O. Estrada, R.C. Genova, B. Girma, E.S. Kissel, A.N. Levy, S. MacCracken, P.R. Mastrandrea and L.L. White (eds.) *Part A: Global and Sectoral Aspects. Contribution of Working Group II to the Fifth Assessment Report of the Intergovernmental*

Panel on Climate Change. Cambridge University Press, Cambridge, United Kingdom and New York, 2014.

Iverson, J. Funding Alaska village relocation caused by climate change and preserving cultural values during relocation. *Seattle Journal for Social Justice*, 12(2), 2013, p. 561–602.

Jiang, X. *Legal Issues for Implementing the Clean Development Mechanism in China*. Google Books, 2013, p. 2.

Jones, N. Sucking carbon out of air. *Nature News*, December 17, 2008.

Jones, N. Sucking it up. *Nature*, 458, April 2009, 1094–7.

Key World Energy Statistics 2014. *IEA*, 2014, p. 6.

Kahn, B. Sea level could rise at least 6 meters. *Scientific American*, July 9, 2015. Available at: http://www.scientificamerican.com/article/sea-level-could-rise-at-least-6-meters/.

Katrina Impacts. *Hurricanes: Science and Society*, 2015. Available at: http://www.hurricanescience.org/history/studies/katrinacase/impacts/.

Kurz, W.A. *et al.* Mountain pine beetle and forest carbon feedback to climate change. *Nature*, 452, 2008, p. 987–990.

Kyoto Protocol. *UNFCCC*, 2014. Available at: http://unfccc.int/kyoto_protocol/items/2830.php.

Kyoto Protocol 10th Anniversary. Timely Reminder Climate Agreements Work. *United Nations Climate Change Newsroom*, February 13, 2015. Available at: http://newsroom.unfccc.int/unfccc-newsroom/kyoto-protocol-10th-anniversary-timely-reminder-climate-agreements-work/.

Kyoto Protocol fast facts. *CNN*, March 31, 2015. Available at: http://www.cnn.com/2013/07/26/world/kyoto-protocol-fast-facts/.

Larsen, J. Record heat wave in Europe takes 35,000 Lives: Far greater losses may lie ahead. *Earth Policy Institute (EPI)*, October 9, 2003.

McMichael, A.J. *et al. Climate Change and Human Health*. World Health Organization, 2003.

Milanovic, B. Global income inequality: A review. *World Economics*, 7(1), 2006, p. 131–157.

National Bureau of Statistics of China, 2015. Available at: http://data.stats.gov.cn/english/.

Oak, R. The evil doers of the financial crisis. *The Economic Populist*, October 16, 2008.

News comes as nations agree negotiating text towards far reaching 2015 Paris agreement. *UN Climate Change Newsroom*, 2015. Available at: http://newsroom.unfccc.int/unfccc-newsroom/kyoto-protocol-10th-anniversary-timely-reminder-climate-agreements-work/.

Now or Never — IEA Energy Technology Perspectives 2008 Shows Pathways to Sustained Economic Growth Based on Clean and Affordable Energy Technology. *IEA*, June 6, 2008.

OECD Ranking of the World's Cities Most Exposed to Coastal Flooding Today and in the Future, extract from OECD Environment Working Paper No. 1. OECD, Paris, 2007. Available at: http://www.oecd.org/dataoecd/16/10/39721444.pdf.

Open Geography, 2015. Available at: http://www.opengeography.org/ch-12-climate-change.html.

Peak oil. *The Economist*, March 5, 2013. Available at: http://www.economist.com/blogs/graphicdetail/2013/03/focus-0.

Pearce, F. Climate warning as Siberia melts. *New Scientist*, August 11, 2005.

Pearce, F. No option left but to suck CO2 out of the air, say IPCC. *Daily News*, April 14, 2014. Available at: https://www.newscientist.com/article/dn25413-no-option-left-but-to-suck-co2-out-of-air-says-ipcc/.

Perry, M.J. Fossil fuels will continue to supply >80% of US energy through 2040, while renewables will play only a minor role. *American Enterprise Institute*, December 16, 2013. Available at: https://www.aei.org/publication/fossil-fuels-will-continue-to-supply-80-of-us-energy-through-2040-while-renewables-will-play-only-a-minor-role/.

Poore, R. *et al.* Sea level and climate. *US Geological Survey (USGS)*, 2012. Available at: http://pubs.usgs.gov/fs/fs2-00/.

Potter, M. The dawn of the green age is delayed. *The Toronto Star*, October 28, 2008.

Pope Francis. Encyclical letter *Laudato si'* of the Holy Father Francis on the care of our common home. *Vatican Press*, May 24, 2015.

Population. *OECD*, 2015. Available at: http://stats.oecd.org/Index.aspx?DatasetCode=POP_FIVE_HIST#.

Poverty Overview. *The World Bank, 2015*. Available at: http://www.worldbank.org/en/topic/poverty/overview.

Preliminary Sigma Estimates: Global Disaster Events Cost Insurers USD 34 billion in 2014. *Swiss Re Economic Research and Consulting*, Sigma No. 1, December 17, 2014.

Preparing for Drought in the 21st Century. *National Drought Policy Commission Report*, 2000. Available at: http://govinfo.library.unt.edu/drought/finalreport/fullreport/pdf/reportfull.pdf.

Press Highlights — 82nd Meeting of the CDM Executive Board. *UNFCCC*, 2014. Available at: https://cdm.unfccc.int/press/newsroom/latestnews/releases/2014/0213_index.html; *UNFCCC*, 2012.

Project Search. *UNFCCC*, 2015(b). Available at: http://cdm.unfccc.int/Projects/projsearch.html.

Rabatel, A. *et al.* Current state of glaciers in the tropical Andes: A multi-century perspective on glacier evolution and climate change. *The Cryosphere*, January 22, 2013. Available at: http://www.the-cryosphere.net/7/81/2013/tc-7-81-2013.pdf.

Ramsey, F.P. A mathematical theory of saving. *Economic Journal*, 38(152), 1928, p. 543–559.

Rodrick, D. Sea Change in the World Economy' article prepared for Techint Conference August 30, 2005, Techint Report 2006.

Ronbine, J.M. *et al.* Death toll exceeded 70,000 in Europe during the summer of 2003; Bangladesh: monsoon floods 2004 — post-flood needs assessment summary report. *Reliefweb*, October 6, 2004.

Sheeran, K.A. (a). Who should abate carbon emissions: A note. *Environmental & Resource Economics*, 35, 2006, p. 89–98.

Sheeran, K.A. (b). Side payments or exemptions: The implications for equitable and efficient climate control. *Eastern Economic Journal*, 32(2), 2006, p. 515–532.

Silicon Valley Index, 2014. Available at: https://www.siliconvalleycf.org/sites/default/files/publications/2014-silicon-valley-index.pdf.

Smith, A. *An Inquiry Into the Nature and Causes of the Wealth of Nations*. Random House, 1776.

Smith, L. Wildlife gives early warning of "deadly dozen" diseases spread by climate change. *Times*, October 8, 2008.

Star Power: The Growing Role of Solar Energy in America. *Environment America Research and Policy Center*, November 20, 2014.

State and Trends of the Carbon Market. *World Bank Group*, 2012, p. 9, 44, 45.

Statistical Review. *BP Statistical Review of World Energy*, 64th Ed., June 2015, p. 5.

Status of Ratification of the Kyoto Protocol. *UNFCCC*, 2015. Available at: http://unfccc.int/kyoto_protocol/status_of_ratification/items/2613.php.

Stern, N. *Stern Review: The Economics of Climate Change. Executive Summary*. Cambridge University Press, 2006. Available at: http://www.wwf.se/source.php/1169157/Stern%20Report_Exec%20Summary.pdf.

Strong growth for renewables expected through to 2030. *Bloomberg*, April 22, 2013. Available at: http://about.bnef.com/press-releases/strong-growth-for-renewables-expected-through-to-2030/.

Sukhdev, P. *The Economics of Ecosystems and Biodiversity*, 2008. Available at: http://ec.europa.eu/environment/nature/biodiversity/economics/index_en.htm.

Swift, T. *A Modest Proposal for preventing the Children of Poor People in Ireland from Being a Burden on their Parents or Country, and for making them Beneficial to the Public*. Sarah Harding, London, 1729.

Swiss Re Economic Research and Consulting, Natural Catastrophes and Manmade Disasters in 2007, Sigma No. 1, 2008.

Swiss Re Economic Research and Consulting, World Insurance 2007: Emerging Markets Leading the Way, Sigma No. 3, 2008.

Swiss Re Economic Research and Consulting. *Sigma Explorer*, 2015. Available at: http://www.sigma-explorer.com.

TEEB for Business Coalition study shows multi trillion dollar natural capital risk underlying urgency of green economy transition. *Trucost*, April 15, 2013.

The Consequences of Global Warming on Weather Patterns. *NRDC*, 2015. Available at: http://www.nrdc.org/globalwarming/fcons/fcons1.asp.

The Deadly Dozen. *Wildlife Conservation Society*, 2008. Available at: http://archive.wcs.org/media/file/DEADLYdozen_screen.pdf.

The Economist, Coal Power Still Powerful. November 15, 2007.

The Emissions Gap Report 2013. United Nations Environment Programme (UNEP), Nairobi. Executive Summary. *UNEP*, 2013, p. xi.

The Kyoto Protocol to the United Nations Framework Convention on Climate Change. *UN*, 1997. Available at: http:// unfccc.int/resource/docs/convkp/kpeng.pdf.

The Times, June 21, 2011.

The World at Six Billion. *UN*, 1999. Available at: http://www.un.org/esa/population/publications/sixbillion/sixbilpart1.pdf.

Thompson, A. Major greenhouse gas reductions needed by 2050: IPCC. *Climate Central*, April, 2014. Available at: http://www.climatecentral.org/news/major-greenhouse-gas-reductions-needed-to-curtail-climate-change-ipcc-17300.

Tollefson, J. Carbon trading market has uncertain future. *Nature*, 2008, p. 508.

Total Carbon Dioxide Emissions from the Consumption of Energy. International Energy Statistics, *EIA*, 2012. Available at http://www.eia.gov/cfapps/ipdbproject/iedindex3.cfm?tid=90&pid=44&aid=8&cid=ww,r6,&syid=2008&eyid=2012&unit=MMTCD.

Total Carbon Dioxide Emissions from the Consumption of Energy. *EIA*, 2014. Available at: http://www.eia.gov/cfapps/ipdbproject/iedindex3.cfm?tid=90&pid=44&aid=8&cid=ww,&syid=2012&eyid=2012&unit=MMTCD.

Tough talks on EU's climate plan. *BBC News*, October 20, 2008.

Trotta, D. Iraq War costs US more than $2 trillion: Study. *Reuters*, March 14, 2013; PollingReport.com, 2015. Available at: http://www.pollingreport.com/iraq.htm.

True cost of coal in South Africa — paying the cost of coal addiction: Executive summary. *Green Peace*, 2010.

Troubled Asset Relief Program (TARP). *U.S. Department of the Treasury*, 2008. Available at: http://www.treasury.gov/initiatives/financial-stability/TARP-Programs/Pages/default.aspx.

Tyndall, J. On the absorption and radiation of heat by gases and vapors and on the physical connexion of radiation absorption and conduction. *Transactions of the Royal Society of London*, 151(2), 1861, pp. 1–36.

Understanding Climate Change: A Beginner's Guide to the UN Framework Convention. *UNFCCC*, July 18, 2000. Available at: http://unfccc.int/cop3/fccc/climate/beginnerg.htm.

UNFCCC Newsroom Press Release, February 13, 2015.

United Nations Environment Program (UNEP). *Global Trends in Sustainable Energy Investment 2008*. UNEP, 2008. Available at: http://sefi.unep.org/english/globaltrends.html.

United Nations Food and Agriculture Organization (FAO). *Livestock's Long Shadow*. FAO, 2006. Available at: http://www.fao.org/docrep/010/a0701e/a0701e00.htm.

United Nations Framework Convention on Climate Change. UN, 1992. Available at: http://www2.onep.go.th/CDM/en/UNFCCCText_Eng.pdf.

United Nations Intergovernmental Panel on Climate Change (IPCC). *Climate Change 2007: Synthesis Report*. In: R.K. Pachauri and A. Reisinger (eds.) *Contribution of Working Groups I, II and III to the Fourth Assessment Report of the Intergovernmental Panel on Climate Change*. IPCC, Geneva, 2007.

United States Census Bureau, 2015. Available at: http://www.census.gov/popclock/.

What's causing the poles to warm faster than the rest of the earth? *NASA*, April 6, 2011. Available at: http://www.nasa.gov/topics/earth/features/warmingpoles.html.

What is the CDM. *UNFCCC CDM*, 2015. Available at: https://cdm.unfccc.int/about/index.html.

Wilkins Ice Shelf News. *National Snow & Ice Data Center*, April 8, 2009. Available at: https://nsidc.org/news/newsroom/wilkins/index.html.

WMO Statement on the Status of the Global Climate in 2014. *World Meteorological Organization (WMO)*. Available at: https://www.wmo.int/media/sites/default/files/1152_en.pdf.

Working Group I: The Scientific Basis. Summary for Policy Makers. *UN IPCC*, 2001. Available at: http://www.ipcc.ch/ipccreports/tar/wg1/index.php?idp=5.

Working Group III — Mitigation of Climate Change. Summary for Policy Makers. *UN IPCC*, 2014. Available at: https://www.

ipcc.ch/pdf/assessment-report/ar5/wg3/drafts/ipcc_wg3_ar5_final-draft_fgd_summary-for-policymakers.pdf.

World Deforestation Decreases, But Remains Alarming in Many Countries. *Food and Agriculture Organization of the United Nations (FAO)*, March 25, 2010.

World Energy Investment Outlook 2014 Factsheet Overview. *World Energy Outlook (WEO)*, 2014.

World Population. *US Census Bureau*, 2015.

World Bank, *State and Trends of the Carbon Market (Annual Report)*, 2006–2014.

World Bank, *World Development Indicators*, 2008.

World Commission on Environment and Development, *Our Common Future* (also known as the Brundtland Report), United Nations, 1987.

Yardley, W. Engulfed by climate change, town seeks lifeline. *New York Times*, May 27, 2007.

Youk, H. and A. van Oudenaarden. Microbiology: Altruistic defense needed for survival. *Nature*, 467(7311), 2010, p. 34.

Zhang, J. *CDM Projects and China's CO2 Emission Reduction*. Royal Institute of Technology, Stockholm, 2010.

Zhang, Z. Is China taking actions to limit greenhouse gas emissions? Past evidence and future prospects. In: W.V. Reid and J. Goldemberg (eds.) *Promoting Development While Limiting Greenhouse Gas Emissions: Trends and Baselines*, UNDP and WRI, New York, 1999.

Glossary of Terms

Adaptive capacity: The capability of a country or region, including on an institutional level, to implement effective adaptation measures.

Allocation: The giving of emissions permits or allowances to greenhouse gas emitters to establish an emissions trading market. The division of permits/allowances can also be done through the grandfathering method and/or auctioning.

Annex I: The term used for the 24 industrialized countries that in 1992 were members of the Organization for Economic Co-operation and Development (OECD) and the 14 countries that at the time were in the middle of a transition from a centrally controlled planned economy to a market economy, including the former Eastern Bloc countries. The European Union (EU) is also in this group. Several countries have subsequently joined, so the group now numbers 41 countries (including the EU).

Annex II: Annex II countries are the same as in Annex I apart from the transition economies. Annex II countries undertake to pay a share of the costs of the developing countries' reductions in emissions.

Annex B: Annex B in the Kyoto Protocol lists those developed countries that have agreed to a commitment to control their greenhouse gas emissions in the period 2008–2012, including those in the OECD, Central and Eastern Europe and the Russian Federation. The latest list of Annex B countries (2007) matches that of Annex I, with the exclusion of Turkey.

The Association of Southeast Asian Nations (ASEAN): Established on August 8, 1967 in Bangkok by the five original Member

Countries: Indonesia, Malaysia, Philippines, Singapore, and Thailand. Brunei Darussalam joined in 1984, Vietnam in 1995, Lao PDR, and Myanmar in 1997 and Cambodia in 1999. The ASEAN Declaration states that the aims and purposes of the Association are: (1) to accelerate economic growth, social progress, and cultural development in the region and (2) to promote regional peace and stability through abiding respect for justice and the rule of law in the relationship among countries in the region and adherence to the principles of the United Nations Charter.

Biofuel: Gas or liquid fuel made from plant material. The source materials can include wood, wood waste, wood liquors, peat, wood sludge, agricultural waste, straw, tires, fish oils, tall oil, sludge waste, waste alcohol, municipal solid waste, and landfill gases. The most common form of biofuel at present is ethanol blended into petrol.

Cap and trade: A cap and trade system is an emissions trading system where total emissions are limited or "capped." The Kyoto Protocol is a cap and trade system in the sense that emissions from Annex B countries are capped and that excess permits might be traded.

Carbon: A basic chemical element in all organic compounds. When combusted, it is transformed into carbon dioxide — one ton of carbon generates about 2.5 tons of carbon dioxide.

Carbon (dioxide) capture and storage (CCS): A process consisting of the separation of carbon dioxide from industrial and energy related sources, transport to a storage location and long-term isolation from the atmosphere (for example, long term storage in oil wells or aquifers).

Carbon dioxide (CO_2): A naturally occurring, colorless, odorless gas that is a normal part of the atmosphere. It is created through respiration, as well in the decay or combustion of animal and vegetable matter. It is also a by-product of burning fossil fuels from fossilized carbon deposits, such as oil, gas and coal, as well as some industrial processes. Because it traps heat radiated by the Earth into the atmosphere, it is called a greenhouse gas, and is a major factor in potential climate change. It is one of the six greenhouse gases that current climate science recommends be cut. Other greenhouse gases are measured in relation to the global warming potential (GWP) of carbon dioxide, and are measured in carbon dioxide equivalent (CO2e).

Carbon neutral: The concept of calculating your emissions, reducing them where possible, and then offsetting the remainder. It also refers to a voluntary market mechanism for addressing CO_2 emissions from sources not yet addressed by climate policy (such as private households, air travel, etc.).

Carbon sequestration: The process whereby carbon dioxide is absorbed in such a manner as to prevent its release into the atmosphere. It can be stored underground (see CCS) or stored in a carbon sink, such as a forest, the soil, or in the ocean.

Carbon tax: A tax by governments, usually on the use of carbon-containing fuels, but can be implemented as a levy on all product associated carbon emissions.

Certified Emission Reduction (CER): A credit or unit equal to one ton of carbon dioxide equivalent, generated under the CDM. The unit is defined in Article 12 of the Kyoto Protocol and may be used by countries listed in Annex I of the Kyoto Protocol towards meeting their binding emissions reduction and limitation commitments.

Chicago Climate Exchange (CCX): The Chicago Climate Exchange is North America's only voluntary, legally binding greenhouse gas (GHG) reduction and trading system for emission sources and offset projects. CCX employs independent verification, includes six GHGs, and has been trading GHG emission allowances since 2003. The companies joining the exchange commit to reducing their aggregate emissions by 6% by 2010.

Clean Development Mechanism (CDM): This is the mechanism laid out in Article 12 of the Kyoto Protocol, that allows Annex I parties to the Kyoto Protocol to make emissions reductions in Annex II countries.

Climate change: This is a term referring to a change in the state of the climate, used to imply a significant change from one climatic condition to another. Note that the UNFCCC, in its Article 1, defines climate change as: "a change of climate which is attributed directly or indirectly to human activity that alters the composition of the global atmosphere and which is in addition to natural climate variability observed over comparable time periods." The UNFCCC thus makes a distinction between climate change attributable to human activities altering the atmospheric composition, and climate variability attributable to natural causes.

Conference of the Parties (COP): The COP is the supreme body of the UNFCCC. Commodity: Something of value that can be bought or sold, usually a product or raw material (lumber, wheat, coffee, metals).

Energy efficiency: The ratio of the useful output of services from an article of industrial equipment to the energy use by such an article; for example, the number of operational hours of a filling machine per kW hour used, or vehicle miles traveled per gallon of fuel (mpg).

European Union (EU): The European Union is an economic and political union of 27 member states, located primarily in Europe. It was established by the Treaty of Maastricht on November 1, 1993 upon the foundations of the pre-existing European Economic Community. The EU has developed a single market through a standardized system of laws that apply in all member states, guaranteeing the freedom of movement of people, goods, services, and capital. It maintains a common trade policy, agricultural, and fisheries policies and a regional development policy.

European Union Emissions Trading Scheme (EU ETS): The EU ETS is the largest multinational emissions trading scheme in the world. It currently covers more than 10,000 installations in the energy and industrial sectors, which are collectively responsible for close to half of the EU's emissions of CO_2 and 40% of its total greenhouse gas emissions.

Framework Convention: Convention that provides a decision-making and organizational framework for the adoption of subsequent complementary agreements (such as a Protocol). Usually contains substantial provisions of a general nature, the details of which can be provided in the subsequent agreements.

Grandfathering: Grandfathering refers to a particular pattern of distributing the rights to pollute. When pollution rights are grandfathered, existing polluters are typically allocated their permits in some proportion to their past emissions levels or emissions activity.

Gross Domestic Product (GDP): The monetary value of all goods and services produced within a nation.

Group of Seven (G7): Now the Group of Eight (G8). See definition below.

Group of Eight (G8): Formerly the Group of Seven (G7). An informal group of some of the world's largest economies, it was originally set up following the oil shocks and global economic recession of the 1970s. The Group consists of France, Germany, Italy, the U.K., Japan, Canada, the U.S., and Russia. The leaders of each country meet every year to discuss how best to manage global economic challenges. The European Union is represented at the G8 by the president of the European Commission and by the leader of the country that holds the EU presidency but it does not officially take part in G8 political discussions. Group of 77 and China (G77/China): The developing country group in the climate negotiations, consisting of more than 130 developing countries.

Host Country: A host country is the country where a JI or CDM project is physically located. A project has to be approved by host country to receive CERs.

Intergovernmental Panel on Climate Change (IPCC): The IPCC was established by the World Meteorological Organization (WMO) and the United Nations Environmental Programme (UNEP) in 1988 to review scientific, technical, and socio-economic information relevant for the understanding of climate change, its potential impacts and options for adaptation and mitigation. It is open to all members of the UN and of WMO. This panel involves over 2,000 of the world's climate experts and the majority of the climate change facts and future predictions covered come from information reviewed by the IPCC.

Joint Initiative: Mechanism of the Kyoto Protocol that allows Annex I parties to the protocol to generate emissions reductions through the development of projects in other developed countries (such as Eastern Europe).

Joint Implementation (JI): A mechanism for transfer of emissions permits from one Annex B country to another.

Keidanren Voluntary Action Plan: An environment action plan devised by the Japan Business Federation aimed at stabilizing

CO_2 emissions from fuel combustion and industrial processes at 1990 levels by 2010. This plan is included in the Kyoto Protocol Target Achievement Plan of Japan, but there is no agreement with the government to assure the targets are reached. The plan makes no commitment to the Japanese government that the target will be met.

Kyoto Protocol: The Kyoto Protocol to the UNFCCC was adopted in 1997 in Kyoto, Japan, at the Third Session of the Conference of the Parties (COP) to the UNFCCC. It contains legally binding commitments, in addition to those included in the UNFCCC. Countries included in Annex B of the Protocol agreed to reduce their anthropogenic greenhouse gas emissions (carbon dioxide, methane, nitrous oxide, hydrofluorocarbons, perfluorocarbons, and sulphur) 5% below 1990 levels in the commitment period 2008 to 2012. The Kyoto Protocol entered into force on February 16, 2005.

Montreal Protocol: The Montreal Protocol on Substances that Deplete the Ozone Layer was adopted in Montreal in 1987 and subsequently adjusted and amended in London (1990), Copenhagen (1992), Vienna (1995), Montreal (1997), and Beijing (1999) . It controls the consumption and production of chlorine and bromine-containing chemicals that destroy stratospheric ozone, such as CFCs, methyl chloroform, carbon tetrachloride, and many others.

New South Wales Greenhouse Gas Abatement Market (NSW): The New South Wales market was the first regulated emissions trading market. New South Wales is Australia's oldest and most populous state. The NSW Greenhouse Gas Abatement Scheme is a state-level program designed to reduce emissions from the energy sector through carbon trading. Under the scheme, NSW energy producers are bound to emit no greater than their share of the NSW per capital emissions target. The target was set at 8.65 tons of CO_2 equivalent in 2003, decreasing by about 3% each year thereafter through to 2007, when it will remain at 7.27. Energy producers exceeding their allotment of emissions can offset them either by surrendering NSW Greenhouse Abatement Certificates (NGACs) purchased from others in the scheme, or by paying an $11/ton fine. Non-Annex I countries: Typically developing countries that have ratified the convention.

North America Free Trade Agreement (NAFTA): The NAFTA was signed in 1992 between the United States, Mexico, and Canada. It is an agreement to remove almost all tariff and non-tariff barriers to trade between the three countries.

Organization for Economic Co-operation and Development (OECD): An international organization of 30 developed countries worldwide. The OECD provides a setting where governments compare policy experiences, seek answers to common problems, identify good practice, and coordinate domestic and international policies.

Regional Greenhouse Gas Initiative (RGGI): The first mandatory, market-based effort in the United States to reduce greenhouse gas emissions. Ten Northeastern and Mid-Atlantic states will cap and then reduce CO_2 emissions from the power sector by 10% by 2018. States will sell emission allowances through auctions and invest proceeds in consumer benefits: energy efficiency, renewable energy, and other clean energy technologies.

Reinsurance: Insurance for insurance companies. It is a way of transferring or "ceding" some of the financial risk insurance companies assume in insuring cars, homes, and businesses to another insurance company, the reinsurer.

Reservoir: A component of the climate system, other than the atmosphere, which has the capacity to store, accumulate or release a substance of concern, e.g., carbon, a greenhouse gas or a precursor. Oceans, soils, and forests are examples of reservoirs of carbon. Pool is an equivalent term (note that the definition of pool often includes the atmosphere). The absolute quantity of substance of concerns, held within a reservoir at a specified time, is called the stock.

Resources: This term refers to the raw materials, supplies, capital equipment, factories, offices, labor, management, and entrepreneurial skills that are used in producing goods and services. It also refers to the substances that support life and fulfil human needs, including air, land, water, minerals, fossil fuels, forests, and sunlight.

Sinks: A sink refers to a pool or reservoir that can absorb and hold onto significant amounts of carbon dioxide. It can refer to the removal of greenhouse gases (GHGs) from the atmosphere through land management and forestry activities that may be subtracted from a country's allowable level of emissions.

Sustainable development: This deals with meeting the needs of the present without compromising the ability of future generations to meet their needs. It is most commonly used in economic terms, to refer to economic development that takes full account of the environmental consequences of economic activity and is based on the use of resources that can be replaced or renewed and therefore are not depleted. For example, environmentally friendly forms of economic growth activities (agriculture, logging, manufacturing, etc.) that allow the continued production of a commodity without damage to the ecosystem (soil, water supplies, biodiversity, or other surrounding resources).

Tipping point: The level of magnitude of a system process at which sudden or rapid change occurs. A point or level at which new properties emerge in an ecological, economic or other system, invalidating predictions based on mathematical relationships that apply at lower levels.

United Nations Framework Convention on Climate Change (UNFCCC): The Convention was adopted on May 9, 1992 in New York and signed at the 1992 Earth Summit in Rio de Janeiro by more than 150 countries and the European Community. Its ultimate objective is the "stabilization of greenhouse gas concentrations in the atmosphere at a level that would prevent dangerous anthropogenic interference with the climate system." It contains commitments for all Parties. Under the Convention, Parties included in Annex I aim to return greenhouse gas emissions not controlled by the Montreal Protocol to 1990 levels by the year.

CPSIA information can be obtained
at www.ICGtesting.com
Printed in the USA
LVHW080852140820
663092LV00002B/13

9 789814 719346